DATA DRIVEN

Data Driven

Truckers, Technology, and the New Workplace Surveillance

Karen Levy

PRINCETON UNIVERSITY PRESS

PRINCETON AND OXFORD

Published by Princeton University Press
41 William Street, Princeton, New Jersey 08540
99 Banbury Road, Oxford OX2 6JX

press.princeton.edu

All Rights Reserved

First paperback printing, 2024
Paperback ISBN 9780691259123

The Library of Congress has cataloged the cloth edition as follows:

Names: Levy, Karen, 1981– author.
Title: Data driven : truckers, technology, and the new workplace surveillance / Karen Levy.
Description: Princeton : Princeton University Press, [2023] | Includes bibliographical references and index.
Identifiers: LCCN 2022021218 (print) | LCCN 2022021219 (ebook) | ISBN 9780691175300 (hardback ; alk. paper) | ISBN 9780691241012 (ebook)
Subjects: LCSH: Trucking—United States—Management. | Supervision of employees—United States. | Truck drivers—United States. | Electronic monitoring in the workplace—United States. | Electronic surveillance—United States. | BISAC: SOCIAL SCIENCE / Privacy & Surveillance (see also POLITICAL SCIENCE / Privacy & Surveillance) | TRANSPORTATION / Automotive / Trucks
Classification: LCC HE5623 .L48 2023 (print) | LCC HE5623 (ebook) | DDC 388.3/240973—dc23/eng/20220624
LC record available at https://lccn.loc.gov/2022021218
LC ebook record available at https://lccn.loc.gov/2022021219

British Library Cataloging-in-Publication Data is available

Editorial: Meagan Levinson and Jacqueline Delaney
Production Editorial: Nathan Carr
Jacket/Cover Design: Karl Spurzem
Production: Erin Suydam
Publicity: Kate Hensley and Kathryn Stevens
Copyeditor: Michele Rosen

This book has been composed in Adobe Text and Gotham

Printed and bound by CPI Group (UK) Ltd, Croydon, CR0 4YY

CONTENTS

DATA DRIVEN

1

Introduction

Ours is a progressively technical civilization [. . .] It is a civilization
committed to the quest for continuously improved means to carelessly
examined ends.

ROBERT K. MERTON[1]

In 2018, a column in *Land Line*, a popular trade magazine for truck drivers,
posed a hypothetical question: "What if the Transformers had ELDs?"[2]
"ELDs" here refers to *electronic logging devices*: digital systems that capture
data about truckers' activities, particularly their work hours, intended to
keep them from driving for more time than federal regulations allow. When
truckers meet the time limits, they are supposed to stop and rest—sometimes
for as long as thirty-four hours—before they can legally drive again.

The columnist, Tyson Fisher, imagines a satirical Michael Bay screenplay in which a squad of Transformers is being summoned by the President
himself to prevent a team of Decepticons from infiltrating Earth. Despite
the urgency and horror of the situation, the Transformers must sheepishly
inform the President that the team cannot help. An excerpt:

> *Optimus Prime and the other Autobots look around the room at
> each other in panic. A sense of helplessness is felt, heard, and seen
> throughout the control room.*
> OPTIMUS PRIME: Ummmmm . . . Mr. President, we're past our hours
> of service. We physically can't move.

FIGURE 1.1. A cartoon imagining the Transformers constrained by electronic monitoring.
Credit: Mo Paul and *Land Line* Magazine.

PRESIDENT (in utter shock): What the hell do you mean??? We're all
    going to die!
OPTIMUS PRIME: It's been over a year since we were needed. Since
    then, Congress passed the ELD mandate. We begged for an
    exemption but never got it [. . .]. There's no wiggle room here.
    We're paralyzed for another 10 hours.
PRESIDENT: Mother of God. What have we done?!?! [3]

The screenplay does not have a happy ending: the Decepticons destroy
the planet and the entire human population, while the Transformers—
hamstrung by digital monitoring and shortsighted regulations—stand by,
unable to save the day.

In December 2017, about six months before the *Land Line* column was
published, the federal government had implemented a requirement that all
truckers buy, install, and use electronic logging devices—the "ELD mandate" to which the column refers. These devices are intended to address one

of the most important and pervasive problems in trucking: fatigue. Truckers are notoriously overworked and underslept, a problem that can have deadly consequences. For decades, truckers have been subject to federal "hours-of-service" regulations that limit the number of hours they can drive each day and each week before taking long breaks. These rules have been in place since the 1930s—and have been enforced, until recently, by requiring truckers to keep track of their hours using paper-and-pencil logbooks, which are subject to inspection by law enforcement at roadside or at weigh stations. But because truckers are economically incentivized to drive for as long as possible, owing in large part to the fact that they are paid by miles driven, it's an open secret in the industry that paper logs are frequently falsified—so much so that classic trucker anthems even make allusions to these regulatory "swindle sheets."

The federal government's solution to the fatigue problem has not been to restructure trucker pay to reduce truckers' incentives to overwork. Rather, it has turned to digital monitoring. ELDs create a digital record of truckers' activities and are intended to be more "tamperproof" and less falsifiable than paper logs (though, as we shall see, they are not completely so). And crucially, these devices facilitate surveillance not only by the government, but also by trucking *firms*—serving as a technological backbone that scaffolds a great deal of real-time data collection about truckers' bodies and behaviors, and fundamentally changing the nature of the trucking workplace.

Though obviously tongue-in-cheek, the *Land Line* Transformers column reflects what many truckers saw as the utter shortsightedness and disrespect of the ELD mandate: they are strong, vigorous, essential workers on whom a great deal (perhaps the fate of humanity!) depends. Yet in their view, they are being hemmed in by a ridiculous device that prevents them from doing their jobs—and we will all pay for it in the long run.[4]

This book examines how truckers' work is being affected by this proliferation of surveillance technologies. These technologies are part of an emerging regime of *digital enforcement*—the use of technology to enforce rules, both legal regulations and organizational directives, more "perfectly" than might otherwise be possible. In trucking, digital enforcement confronts the existing social order of the industry: it upends the occupational autonomy truckers have traditionally held, reconfigures information flows within trucking firms, alters how truckers and law enforcement officers interact with one another, and creates new sites of contestation and resistance. The economic realities of trucking have long depended on truckers' discretion, including flexible recordkeeping routines and the ability to direct their own work in the face of unpredictable and often inhospitable conditions. But digital enforcement doesn't address these realities; it papers over them. In this book, I examine the social consequences of this approach.

Are Rules Made to Be Broken?

The rationale behind the ELD mandate is that "electronic logs take the non-compliance issues off the table" (as one trucking executive put it),[5] making it much less feasible for truckers to break the hours-of-service rules. We might, of course, accept this goal as normatively desirable and relatively uncontroversial. If we have rules in place for safety reasons, at which we have arrived through legitimate political processes, and those rules are easily broken, why *shouldn't* we enforce them more consistently if we can do so? Aren't rules rules?

In fact, it's not so simple—rules *aren't* always rules. Rules are shaped by social, cultural, and economic realities, and are almost never as simple as they might seem on paper. As an intuitive example, consider how you'd feel if you were ticketed by a police officer for driving sixty-six miles per hour when the speed limit is sixty-five. Officially, you've broken the rule—but the violation is so trivial, and the rule is so commonly broken, that to-the-letter enforcement would very likely strike us as unfair and unreasonable. (Further proving the point, the policing of minor speed violations is often merely a "pretext"—that is, a disingenuous excuse—for police to stop and investigate drivers, particularly Black drivers, for other reasons.[6])

But there's an even more important point here. It's not just that some amount of rule-breaking, like the behavior of our barely speeding driver, is considered unobjectionable. The more fundamental point is that we often *depend* on rule-breaking to make the world function. When former New York City Mayor Bill de Blasio proposed a zero-tolerance policy for jaywalking, economists were quick to point out that if anti-jaywalking rules were followed to the letter, pedestrian commute times would increase, and the city's social and economic life would suffer for it.[7] We rely on this sort of routine rule-breaking to make society function efficiently. Perhaps no phenomenon illustrates the point more clearly than the "work-to-rule" labor action, in which unions exert pressure on management by following every rule in the handbook to the letter. Doing so slows organizational functions to a halt—but because employees are *officially* following the rules, it is difficult to discipline them. The fact that working to rule functions as a resistance tactic demonstrates by implication that most of the time, work practices *do not* fully accord with the rules—indeed, organizations rely on the fact that they don't.[8]

Examples abound—particularly in the workplace. Organizational theorists have long observed that firms, in response to pressures for institutional conformity, often adopt certain formal rules and structures to achieve legitimacy, while decoupling those rules from practices that allow them to function efficiently.[9] Alvin Gouldner's seminal 1954 ethnography of a gypsum mine describes workplace rules that were in neither managers' nor workers' interests to enforce.[10] These rules were enforced only on rare occasions—for instance, a smoking ban was enforced only when insurance inspectors came by—while most of the time managers "looked the other way" at employee rule-breaking. This "mock bureaucracy" effectively gave supervisors a bargaining chip they used to maintain friendly relationships with workers. More recently, Michel Anteby described similar practices in an aeronautical factory, where managers tolerated and even encouraged employees making small souvenirs, called "homers," for new retirees, using company time and materials.[11] Though homer-making was officially forbidden, the fact that this rule was so often broken served a similar function as Gouldner's smoking ban: managers could ultimately exert more control over workers by seeming to be "on their side" when it came to rule-breaking, and workers viewed the practice as a source of pride and occupational identity.

To be sure, nonenforcement of rules, in the workplace or more generally, isn't necessarily a good thing. Many rules are in place for a good reason—to protect workers and others in vulnerable positions, or as a check on the powerful. Selective enforcement of laws can be the basis of arbitrary or discriminatory treatment. And we may or may not think that the managers in Gouldner's gypsum mine and Anteby's aeronautical plant were ultimately doing workers any favors when it came to smoking and homer-making: "looking the other way" for certain rule violations gives managers leverage to pressure workers in other ways, and managers can "play favorites" by enforcing rules in some situations (or against some workers) and not others.[12]

But my goal here is not to argue that strict, to-the-letter enforcement of rules (or nonenforcement on the other hand) is altogether "good" or "bad"—the world is far too complex to make such a sweeping statement. Instead, I raise these examples to illustrate *how much of social life*, in the workplace and more broadly, relies on the "gap" between rules on the books and practices on the ground.[13] The "wiggle room" around rules is a site for strategic negotiation, for economic functioning, for relationship management—for both good and ill. So when we decide to more strictly enforce rules using technology, without accounting for what has been happening in the gap, we may well disrupt the social order of a particular context in important and unforeseen ways.

Despite this, the notion of "more perfect" enforcement via digital technology is a common refrain, often motivated by the idea that technology can help us to close the gap between rule and practice—transforming society into a more consistent and just version of itself. As David Friedberg, then CEO of data analytics firm Climate Corporation, described it in 2014: "Technology is the empowerment of more truth, and fewer things taken on faith."[14] Using technology, the logic goes, we can ascertain what people *really* do, instead of what they say they do; we can catch and deter cheaters and liars; we can generate knowledge where before we had only hunches and secrets; we can make people follow the rules.

Sometimes, digital enforcement happens through attempts to *prevent* violation, making rules more difficult to break—using code to make it more onerous (or even impossible) to deviate from an imposed rule.[15] For example, digital rights management technology makes it (nearly) impossible to violate copyright law.[16] If these technologies work as "perfectly" as intended, rule violation is completely impaired, and violation becomes practically impossible (or at least much more difficult).[17] But even more common than tools of prevention are tools of *detection*—technologies that function not by making rule-violating behavior more difficult to execute, but by creating a comprehensive *account* of our behaviors. These are surveillance technologies. For example, body-worn cameras don't make it impossible for a police officer to use unauthorized force against a civilian, but are intended to make the officer more accountable should they do so.[18] These technologies may work by deterring sanctioned behaviors—knowing that one is being observed can incentivize rule-following—or because they enable enforcers to more swiftly detect and punish rule-breaking.

Perhaps nowhere do we see this trend more clearly than in the workplace, where surveillance over workers' behaviors has become a favored method for compelling compliance with the aims of management. As we'll see, this practice has deep roots—but contemporary workplace surveillance has some new features, too.

Work and the Future of Work

We often anticipate the "future of work" in either dreamy or dystopian terms. The phrase has been widely adopted by technologists and commentators, either to describe a paradisical ideal in which people have much greater autonomy and flexibility to do work in ways that suit them while affording them ample time for leisure; or, as a dark alternative, as a future in which workers have ever-diminishing social and economic power and in which their every move and thought is overseen, predicted, and optimized by management, human or algorithmic. Both visions, though, are united by the assumption that the future of work (whatever it looks like) will happen, well, in the future—that is to say, this is a *vision* of a time that is not now, and that is somehow different than now, or at least different enough that it deserves its own label.

It's rather curious that we tend to talk in such future-oriented prognostications about what technological change will portend for work and the workplace. In other domains, the way we talk about technology tends to be more focused on what is occurring now or in the very near term; but when it comes to work, we maintain some temporal distance, at least in our discourse, from these changes. This is odd because the "future of work" is, of course, not some distant or discrete mode of social organization so unlike the one we have today. The management practices of tomorrow are, in many ways, not particularly different from the management practices of the past. They're built on the same foundations—motivating efficiency, minimizing loss, optimizing processes, improving productivity.[19] And one of the most common strategies for achieving these goals, then and now, is increased oversight over the activities of workers.[20]

So what *is* new about today's workplace surveillance? Is this not just more of the same, driven by the same organizational goals that have always motivated managerial oversight—even if the specific technologies that are used to do so have changed form in one way or another? Some workplace monitoring *is* old wine in a new bottle, a contemporary instantiation of the manager with a clipboard looming above the factory floor. This is not to say, of course, that these practices don't deserve scrutiny or critique—but we should be precise about what, if anything, is new here.

In fact, there are some subtle but important dynamics that distinguish contemporary workplace surveillance from what's come before, and that will become important in telling the truckers' story. First, contemporary technologies facilitate surveillance in *new kinds of workplaces*. Geographically distributed and mobile workers, for example, have historically maintained more independence from oversight than workers centralized in nonmobile workplaces, like factories, call centers, and office buildings—but location tracking, sensor technology, and wireless networking have changed that. Porous boundaries between home and work also facilitate surveillance in new places. For example, the growth of work-from-home arrangements during the Covid-19 pandemic has led to greater use of tracking software to monitor workers' keystrokes, locations, and web traffic—as well as video capture of the kitchen tables and living rooms in which their work now takes place.[21]

New kinds of data also come to the fore. As sensor technologies become cheaper and easier to deploy, and workplace surveillance capability is more frequently embedded in software by default, employers are well positioned to capture more and more fine-grained data about workers' movements and activities. Wearable technologies, like those used in Amazon's warehouses, monitor and evaluate workers' speed with much more precision than was previously possible—including the number and length of their bathroom breaks.[22] Employers increasingly monitor and analyze datapoints like workers' social media posts, phone calls, and attendance at meetings; Microsoft faced pushback in 2020 when it built "productivity scoring" into its widely used Office 365 product, which gave managers access to "73 pieces of granular data about worker behavior" like email and chat frequency.[23] And as we'll discuss, biometric data is also becoming more commonly collected in the workplace—from authentication mechanisms like fingerprint and retinal scans to behavioral data about workers' attention and fatigue.

These new data streams fuel *new kinds of analysis* that impact how workers are managed. In some contexts, managerial decisions are implemented through opaque algorithmic systems that can create acute information asymmetries between workers and firms—like Uber's use of algorithms to apportion rides and determine rates without making those rules transparent to its drivers.[24] Other analyses are predictive, designed to forecast which workers are likely to be most productive, how many workers to staff at a given time to meet demand, or which worker is likely to make a sale to a particular customer.[25]

Finally, contemporary workplace surveillance can blur boundaries between the workplace and other spheres of life, creating *new kinds of entanglements* across previously disparate domains. Surveillance of work-from-home environments can facilitate data collection about family, friends, and living situations. Managers often keep tabs on workers' online activities on social media platforms.[26] Workplace wellness programs can facilitate employers' collection of data about worker health, stoking concerns about discrimination.[27] And "bring-your-own-device" policies, in which an employer's software is installed on a worker's own personal phone or computer, can further muddle distinctions between home and work and create additional data privacy and security concerns.[28]

The Trucking Workplace

All of these dynamics are at play in trucking. For decades, the mobile, isolated nature of truckers' labor provided a buffer against managerial oversight, giving truckers much more freedom and autonomy in their day-to-day work than other blue-collar workers. But this has changed drastically, owing to the proliferation of digital monitoring technologies like the ELD: the road no longer affords drivers the independence they once had. As we'll see, surveillance in trucking involves the collection of new kinds of fine-grained data about truckers' behaviors, bodies, braking patterns, even brainwaves. This data collection supports new forms of analysis: firms can compare truckers' performance against one another and predict what they are likely to do in the future—giving them much more visibility into, and control over, truckers' work than ever before. And as this book will explain, trucker surveillance involves deep entanglements among the interests of many different actors in different social spheres—not only the trucker, the government, and the firm, but also truckers' families, insurers, third-party companies seeking to make money from the data, and the public at large. As such, truckers may be canaries in the coal mine: investigating digital surveillance and rule enforcement in this industry can give us important clues about how these dynamics may function in other contexts, both within and outside the workplace.

Trucking is a job, but it is more than that. Trucks are understood by their drivers both as workplaces, relatively free of meddlesome bureaucratic oversight, and as homes, in which they live, eat, and sleep for days or even weeks at a time, and in which their privacy is sacrosanct. And similarly, understanding trucking as the mere activity of driving a truck is only one facet of what trucking means to those who call themselves truckers. Trucking work is bound up with cultural constructs of manhood and virility, performed through displays of physical and mental stamina. The icon of the "asphalt cowboy" has been an iconic figure in trucking for decades;[29] workers who strain against authority in traditional employment settings may self-select into trucking as an occupation in which the day-to-day routines of work have, traditionally, been largely self-directed. Trucking is an *identity*: an enactment of masculinity, a form of economic provision, and an extension of sexuality. Sociologist Raewyn Connell wrote that for working-class men, "bodily capacities *are* their economic asset"[30]—the ability to push their bodies up to and past the limit is how they maintain economic autonomy. The ELD reduces this autonomy and impugns the self-knowledge on which it relies. To a far greater extent than in many other workplaces in which digital monitoring has proliferated, in trucking such monitoring clashes acutely with truckers' collective and individual self-definition and occupational identities, built up over decades.

The ELD, then, operates simultaneously as a legal, economic, and cultural object. To regulators, it's a legal creation—the product of federal regulations designed to enforce compliance with rules, and a strategy to address safety problems that plague the industry (though with mixed evidence of effectiveness). To firms, it's primarily an economic tool to align workers' behaviors with organizational aims such as maximizing fuel efficiency and minimizing out-of-route driving. And trucking firms are major stakeholders in the formulation of the legal rules, since they too have interests in how much their drivers (and their competitors' drivers) may legally drive. To truckers, the ELD is a cultural object that challenges the value of their "road knowledge" and occupational identity, as well as long-held industry norms for getting work done independently. In this book, I'll consider how the ELD works in each of these ways, and how these different meanings of the ELD interact and clash with one another. The complex interplay of legal rules, socioeconomic organization, cultural norms, and technical capabilities that we find in trucking makes the industry a strategic site[31] for examining the interaction of multiple domains in shaping surveillance technologies.[32] Only by examining these dimensions together can we understand how and why digital enforcement works—or doesn't.

There are also key pragmatic reasons why trucking offers a strategic site for investigating digital enforcement and workplace surveillance. I examined trucking during a period of technological transition. To understand the impact of the ELD on the industry, I conducted ethnographic research with truckers, trucking managers, regulators, technologists, and others affiliated with the industry, across a wide variety of sites—trucking firms, truck stop bars, trade shows, regulatory meetings, and more. The bulk of my fieldwork occurred between 2011 and 2014—when the ELD was gaining traction within the industry and was the subject of active debate, but was not yet mandatory—and was supplemented through 2019, after the mandate had taken effect. (In Appendix A, I offer much more detail on the contours of the study, including my data sources and methodological approaches.) Studying ELDs during this period of transition was essential because it allowed me to observe and analyze differences between analog and digital systems. Moments in which the status quo is disrupted can be particularly illuminating occasions for unearthing the entrenched assumptions and power arrangements that undergird social organization.[33] Of course, ELD use was far from randomly distributed in the industry during this period; for reasons I'll discuss in chapter 3, large carriers tended to adopt the devices much earlier than small firms did. Nonetheless, the transition period was crucial to understanding and comparing practices with and without digital enforcement (as I do, for instance, in chapters 4 and 5) to fully understand its impacts on the industry.

The Road Ahead

This book does three things. In the first part of the book (chapters 1–3), I describe the problem that electronic surveillance is, ostensibly, being marshaled to solve in trucking—and what kind of political, economic, and cultural environment it is being integrated into. My goal in this section is to illustrate why using digital technology to enforce the rules in this environment is unlikely to be a successful endeavor; the social and economic history of trucking is crucial to understanding why.

Trucking is a kind of workplace unlike any other. Chapter 2 delves into the fascinating, and contentious, political and economic history of the long-haul trucking industry in the United States. Truckers were viewed as respected "knights of the highway" in the mid-twentieth century. But their professional public image had been decimated by the 1980s, when economic deregulation created a race to the bottom that depressed truckers' wages, made their work significantly more difficult, and led many truckers to feel like "throwaway people." The workday is long, dangerous, and extraordinarily tiring, and wages are significantly lower than they once were. As a result, the industry suffers from an incredible amount of labor churn, in which drivers enter and exit jobs quickly—either hopping among companies in search of more favorable conditions or exiting the industry entirely. Rather than responding by making conditions more humane or wages more competitive, the industry has responded by trying to compel new pools of potential drivers to enter trucking—for example, newly licensed eighteen-year-olds—often to the detriment of public safety. I discuss the peculiar per-mile pay structure in the industry, and truckers' legal exemption from federal labor standards—conditions that keep drivers from being paid for the true amount of work they do and that incentivize dangerous driving and overwork. Yet, despite all the economic and social problems plaguing the industry, trucker culture is strong and proud. Truckers' professional identity has long celebrated their independence, biophysical stamina, and control over their day-to-day work—even in the face of tightening controls from federal regulators and companies.

Electronic surveillance enters the picture in chapter 3. In this chapter, I explain why truckers are so tired, what the government has tried to do about the problem, and how the ELD mandate eventually became the favored solution. By and large, truckers' reception of ELDs was not positive: truckers invoked concerns about their autonomy and privacy, as well as how more stringent enforcement of the hours-of-service rules was likely to harm them economically and to exacerbate logistical issues. In short, strict digital enforcement is at odds not only with the way the industry actually functions, but with its culture. We might accept these as necessary trade-offs if they in fact increased the safety of the public roads—but the best evidence we have demonstrates that the safety outcomes of electronic surveillance are mixed at best, and may actually lead to *more* accidents because of the inflexibility it introduces into trucking work.

Digital monitoring entered trucking as a technology of regulation—but its impacts on the industry have extended far beyond government enforcement of timekeeping rules. The second goal of this book is to show how digital monitoring has wide-ranging and complicated implications for social and organizational relationships in the industry. In the second part of the book (chapters 4–6), I take deep dives into understanding the ELD's role in three overlapping social spheres: as a tool for business, an object of inspection, and a locus for resistance.

I begin, in chapter 4, by investigating how the ELD operates as a business tool. The ELD augments trucking firms' ability to manage drivers in new ways—replacing the value of truckers' local and biophysical "road knowledge" with abstracted, easily compared data flows. The ELD is very often bundled with other monitoring capacities that track, in fine detail, many aspects of a trucker's behavior—how hard he brakes, how much fuel he uses, whether he is paying attention to the road—and gives back-office managers real-time access to all kinds of information about a trucker's day-to-day work, information that used to be within the exclusive purview of the trucker. Firms use this knowledge in a variety of ways—using analytics to foster competition among drivers, even leveraging drivers' connections with their families—in order to compel them to act in ways that favor the firm's goals. Even beyond managerial use, data from ELDs become valuable to other business interests—including insurers, freight brokers, and others; while some of these secondary uses might favor truckers, others may impose additional burdens on them. Further, monitoring systems play a role in the competitive political economy of the industry, offering comparative advantages to large firms while imposing disproportionate costs on independent truckers and mom-and-pop operations. We can think of government, corporate, and third-party data collection as being *interoperable*; while each has its own aims, the ELD bundles them together into a mutually enforcing whole.

Chapter 5 looks closely at the role ELDs played in inspection interactions between truckers and law enforcement officers prior to the ELD mandate. In the same way that ELDs have made some aspects of truckers' work more difficult, they similarly disrupted the work of commercial vehicle inspectors, who were unused to their inconsistent and often confusing interfaces. Inspectors are charged with ensuring the accuracy of a trucker's logbook, and they have cultivated their own professional practices for doing so. ELDs upended these practices in important ways. Inspectors inspecting paper logs once made strategic use of time and distance to establish authority—separating driver from log to conduct a careful, self-paced inspection of the driver's records. But because ELDs are often fixed *within* truck cabs, inspecting ELDs put the inspector in the cab *with* the driver, and put his lack of expertise about dealing with the ELD in the driver's full view. As a result, inspections of electronic logs were socially uncomfortable, undermined the inspector's authority, and could be far more cursory than inspections of paper logs—perhaps leading to some of the reduction in citation rates for drivers with ELDs, and counterintuitively placing the trucker and inspector into temporary alignment against a digital "common enemy." What's more, because truckers knew about inspectors' hesitation to inspect ELDs, they engaged in *decoy compliance*—strategically signaling to inspectors that they had an ELD in the cab, perhaps without actually having one—in order to convince the inspector to wave the truck through the weigh station without inspecting it. These practices show us that digital monitoring doesn't *necessarily* reinforce the authority of the state over the surveilled subject. The interactions that real people have around these technologies complicate this dynamic considerably and show us how these technologies can also destabilize power relations in surprising ways.

In chapter 6, I turn to how truckers "beat the box"—the wide variety of strategies truckers use to resist digital monitoring. These strategies run the gamut from smashing the monitor with a hammer to organizing collective protests; from exploiting an ELD's technical limitations in order to eke out more driving time to "hacking" the device to play computer solitaire. These strategies make use of different resources, involve different parties (sometimes including trucking companies themselves!), and have different goals—while some are designed to communicate, others are designed to obfuscate; while some are aimed at the government, others are aimed at companies; while some are about creating the flexibility to earn a living, others are about reasserting one's identity. The variety of trucker resistance practices raises important questions about what—and whom—resistance is for: does resistance "count" if it doesn't change anything structurally, or even reinforces exploitative labor conditions? Is resistance a "weapon of the weak," or can it be a way to push the least powerful even further down the ladder? The answer is that resistance to surveillance is all these things at once, and we understand resistance more clearly if we think of it as a means of negotiating social power, rather than merely as bottom-up refusal.

Finally, in the third part of the book (chapters 7 and 8), I look to the future. Here, my goal is to connect our study of electronic monitoring to other dynamics—the relationship between surveillance and automation, and the role surveillance may play in enforcing rules more generally. Our case study of the ELD provides a window to help us forecast what kinds of obstacles emerging technologies are likely to face and some ideas about how to think about technology as a tool to address social problems.

Chapter 7 asks: what will trucking look like in the near future, as autonomous vehicles become more widespread? Will there even be human truckers in twenty years, or will robots have taken over these jobs? Several economic forecasts have described trucking as a prime target for automation as artificial intelligence (AI) becomes more and more capable. And we can understand why: autonomous trucks don't get tired and might do a lot of the trucking labor that is so difficult and dangerous for humans; and they could, in theory, save millions of dollars in labor costs for trucking firms. The potential for human truckers to be displaced from the truck cab, physically and economically, raises serious concerns. But the reality is that autonomous technologies are very unlikely to replace truckers in one fell swoop, for a variety of legal, social, and cultural reasons. Instead of thinking about a cliff of sudden job loss, we should think about a gradual *slope*, in which humans and machines integrate their work with one another. In this chapter I consider various scenarios describing what that integration between human truckers and autonomous trucking might look like. I examine some of the technical and organizational roadblocks faced by different models of integration, and explain that the use of artificial intelligence in the trucking industry today is a very different experience than the displacement often imagined and feared. Rather than being kicked out of the truck cab by technology, the trucker is still very much in the cab, doing the work of truck driving— but he is increasingly *joined there* by intelligent systems that monitor his body directly and intrusively, through wearable devices and cameras, often integrated into the fleet management systems we discussed in chapter 4. AI in trucking is experienced as a *hybridization* of man and machine. Surveillance and automation in trucking are complements, not substitutes. In order to assess the true range of potential effects of AI on trucking work, we need to consider both the possibility of job replacement *by* robots and of bodily integration *with* robots.

To conclude, in chapter 8, I return to the issue of using technology to enforce rules, both within and beyond the workplace. Technology often fails as a solution because the problems it's intended to solve aren't, at their core, technology problems—they're social, economic, and cultural problems, and they require solutions in the same register. Trying to address these problems via technology often means insufficiently accounting for the world as it is. But this doesn't mean that technology has *no* role to play in addressing social problems, and I offer a few possibilities for how we might think about using technical solutions in the service of broader reforms.

2

If the Wheel Ain't Turnin', You Ain't Earnin'

TRUCKER POLITICS, ECONOMICS, AND CULTURE

> I'm a prisoner of the highway
> Driven on by my restless soul
> Call me a prisoner of the highway
> Imprisoned by the freedom of the road.
> **RONNIE MILSAP, "PRISONER OF THE HIGHWAY" (1984)**

To appreciate how surveillance is transforming the trucking workplace, we first need to understand what kind of workplace trucking is. We need to know how trucking operations are organized; what economic pressures bear on trucking companies, and on drivers; and who truckers conceive themselves to be culturally. With this context in mind, we will be better equipped to understand how and why new monitoring technologies are being integrated into trucking work—and what meanings they take on when they do so.

If You Got It, a Truck Brought It

It's hard to overstate how fundamentally reliant the United States economy is on the trucking industry. Trucks, truck routes, and truckers themselves constitute the critical infrastructure upon which markets for essentially every tangible good in the United States depends. A well-loved adage in the industry is "if you bought it, a truck brought it": even in the digital age, nearly every object each of us touches has been—at some point, in some form—on a truck, being driven by a truck driver, across miles of highway. At the time of this writing, supply chain problems stemming from the Covid-19 pandemic are disrupting the production and distribution of goods world-wide. As is often the case, we only appreciate just how crucial the labor of truckers is, and how profoundly we have come to depend on them for our world to function, during such moments of crisis.

The scale of the industry is staggering. In 2018, thirty-seven million trucks were registered for business purposes in the United States, and together drove an estimated 305 billion miles that year, transporting nearly twelve billion tons of freight.[1] And the volume of goods transported by truck is only expected to increase. Heightened expectations of fast, customer-trackable shipping, from Amazon Prime and the like, increase the urgency with which trucks (and truckers) are expected to haul goods across the country.

The number of people involved in this endeavor is similarly enormous. Nearly eight million people work in the trucking industry (as of 2018); of these, about 3.6 million are truck drivers,[2] and about two million of these drive "the long haul."[3] Long-haul trucking—the focus of this book—involves drivers routinely driving very long distances, far from their home base for days or weeks at a time, to move goods across the country.

But despite the enormity of the trucking industry and the dependence of the economy on trucking, we tend not to notice or think much about it in our day-to-day lives—at least when supply chains are functioning normally. Truckers' work, though it takes place almost entirely in public, is relatively invisible unless something goes wrong. When we *do* notice truckers on the highway, it's often with annoyance or trepidation—say, when we're trying to pass them on a crowded turnpike—or even with fear. Portrayals of truck-ers in popular culture tend to be stunningly negative and associated with confrontation, illegality, or even violence. The correlation of trucking with violence in the American imagination stems, at least in part, from some high-profile investigations of truck-driving serial murderers in the 1990s and 2000s. In 2004, the FBI created the Highway Serial Killings Initiative, a program intended to link unsolved murders (often of sex workers) with suspicious activity by long-haul truckers. While credited with helping law enforcement to solve some outstanding cases, the initiative did little to bur-nish the public reputation of the American trucker.[4]

Across America, communities pass ordinances to keep trucks and truckers out of their communities. They ban trucks from using their local roads and prevent truck parking from being built at rest stops—despite the fact that truck parking is essential for truckers to get enough (legally and biologically) necessary rest. Even though we depend on truckers to keep the economy running and to get our deliveries on time, a strong strain of NIMBYism ("not in my back yard!") emerges when, as a public, we have to actually provide infrastructure for truckers' needs. As one veteran trucker interviewed by the *New York Times* in 2020 put it: "We're throwaway people. Nobody cares about us. Everybody's perception of a truck driver is we clog up traffic, we get in the way, we pollute the environment. . . . Everybody needs us, but nobody wants us."[5]

The Knights of the Highway

It wasn't always this way. Truckers used to be hailed as heroes of the highway—the safest and most reliable drivers, doing honorable, skilled, and important work. Some of this respect stemmed from truckers' expertise with the mechanics of driving at a time when such knowledge was less commonplace. A 1938 promotional film on safe night driving, produced by Chevrolet, heralded long-haul truck drivers as "true knights of the highway in every sense of the word":

> These truckers go out of their way to help others on the road. The driving public has come to know them as the most courteous drivers on the highway. And with all of this, they have established some of the finest safety records in the country. . . . For dependability, endurance, courtesy, and skill, we will look far before we find a class of drivers who handle themselves and their vehicles with greater safety than the men behind the wheel of the big interstate trucks.[6]

Because cross-country truckers did much of their driving at night, they were seen as experts on the nighttime driving practices to which the newly motoring public needed to be introduced to maintain the safety of the public roads—dimming one's headlights while passing an oncoming vehicle, for instance. Trucking was treated as a *profession*, not just a job—and accorded the respect that came with recognition of truckers' professional expertise.

FIGURE 2.1. "Knights on the Highway," a promotional film produced by the Jam Handy Organization, 1938, archived at https://archive.org/details/Knightso1938.

FIGURE 2.2. Image from 1952 American Trucking Associations promotional booklet "Heroes of the Highway." Reproduced in "The Truck Driver, Comic Book Hero," *Wheels of Time* 32, no. 5 (September/October 2011), 14.

Racing to the Bottom

The demise of truckers' public image from hero to scofflaw accompanied major structural changes in how the mammoth trucking industry operated. For most of American history, trucking was an economically regulated industry: prices were standardized, and market structure was dictated largely by federal regulations. The passage of the Motor Carrier Act of 1935 brought trucking firms under the regulatory purview of the Interstate Commerce Commission, or ICC. The ICC set standard rates within the industry and created protective barriers against new market entrants, shielding existing companies from competition. Truckers who wanted to operate on the public roads were required to obtain scarce "certificates of public convenience and necessity" in order to do so, which made it difficult for new companies to establish a foothold in the industry.

In the 1970s, trucking was one of a few key industries targeted by the "deregulatory wave," a political reform movement that pushed for increased competition and reduced federal involvement in determining rates and working conditions. Deregulation was premised on the notion that heavy regulation was economically inefficient and resulted in excess costs being pushed to consumers. Trucking deregulation began in earnest in 1977, when new laws began to chip away at the ICC's authority and dismantle several of the regulatory structures surrounding the industry. Deregulation was formally enacted by the Motor Carrier Act of 1980, which permitted free market entry and allowed carriers to charge discount rates.

These moves had swift and significant effects: By 1982, shipping rates had declined by 25 percent, and service quality improved in light of the newly competitive climate.[7] But along with these benefits for consumers came decidedly negative effects for the frontline workers directly affected by deregulation: truckers. Thousands of truckers lost their jobs in the years immediately after deregulation took effect, as trucking companies cut back on costs or were put out of business altogether. The economic recessions of the early 1980s didn't help matters.[8]

Among those who kept on trucking, working conditions changed drastically. As the financial margins on which trucking companies operated became thinner, truckers worked more hours and for less money. During the ten years immediately following the advent of deregulation (from 1977 to 1987), truck drivers' wages dropped by an astonishing *44 percent*, forcing truckers to drive much longer hours—and, as we will see, often to break the law—to make economic ends meet.[9] The industry fared far worse than other blue-collar jobs. Between 1977 and 1995, the decline in truck drivers' average real earnings was *four times* that of demographically comparable workers in manufacturing production.[10] Deregulation was devastating to truckers' paychecks, and truckers' economic lot has not significantly improved since then. In 1980, truckers made roughly $110,000 in today's dollars.[11] In contrast, as of 2020, truckers' median annual earnings have stagnated at about $47,000 per year—a figure that has not budged significantly for about fifteen years.[12]

The "race to the bottom" engendered by deregulation took its toll on drivers not only economically, but in terms of the conditions under which they did their day-to-day work. Overwork and cost cutting wreaked havoc on drivers' physical and mental health and well-being—leading transport economist Michael Belzer to compare post-deregulation trucking work-places to "sweatshops on wheels." In Belzer's analysis, the combination of low wages, overwork, and unsafe conditions in trucking made trucking comparable to the infamously poor labor conditions in sweatshops.[13] Truckers had always worked hard, but newly unfettered competition pushed them even further past their limits. As drivers worked longer hours for less pay, they dealt ever more with the strains of life on the road, with unsafe and unhealthy working conditions, with isolation and family instability wrought by the demands of the trucking life, and with the grave dangers of their occupation. Truckers are far more likely to be killed or injured on the job than the average American worker. One out of every six people killed on the job in the United States is a truck driver[14]—and the number has been steadily increasing to now-record levels,[15] despite the addition of new safety features. As of 2019, trucking had the sixth highest occupational fatality rate of all American jobs;[16] the incidence rate for nonfatal but serious injury among truckers is the highest of any US occupation, and three times that of the average worker.[17]

Even beyond accidents, trucking is a physically and mentally grueling line of work in almost every conceivable way. Truckers have a very difficult time eating nutritiously at truck stops or getting regular exercise; they regularly face severe health consequences caused or exacerbated by living on the road, ranging from exhaustion to addiction, from chronic pain to repetitive stress injuries, from risky sexual behaviors to post-traumatic stress disorders after witnessing gruesome accidents.[18] They suffer higher-than-average rates of diabetes, spinal problems, cardiovascular disease, hypertension, depression, and myriad other conditions; truckers have the highest incidence of obesity of any group of workers in the United States.[19] What's more, truckers often have no access to physical or mental health services—they can rarely attend appointments in person, and often lack enough control over their own schedules to know when they might be home to visit a doctor, dentist, or therapist.

As a result of all these difficulties, today's trucking labor market is marked by significant churn. Driver turnover in the industry hovers, remarkably, close to *100 percent per year* for large firms:[20] that is, on average, nearly every trucker will leave their employer every year (the rate is somewhat lower, but still quite high, for smaller firms). Multiple factors contribute to the astronomical rate of turnover in the industry, but the harsh working conditions the job entails in light of its low wages and overwork are undoubtedly high on the list. Truckers also frequently hop between firms, seeking better wages or more desirable hauls; comparing tips on what different companies are paying or how many hauls one can get is a common topic of small talk at truck stops. Another major contributor to driver turnover is the fact that the trucking population is aging rapidly: the median trucker is 46 years old.[21] And the share of younger employees in the industry has been *decreasing*: between 1994 and 2013, the segment of the trucking labor force older than 55 grew by 126 percent, while the segment between 25 and 34 years old declined by almost half. The industry's reliance on an aging cohort of drivers—what the American Transportation Research Institute calls the "trucking generation"—spells trouble for the industry in the coming years.[22]

The aging of the labor force and the astronomical rate of turnover contribute to what's often referred to as trucking's labor shortage: firms are constantly recruiting new workers to meet the demand of moving freight quickly. In 2018, the industry had about fifty thousand unfilled jobs, creating bottlenecks to timely delivery;[23] by 2021, the shortage had ballooned to a historic high of eighty thousand unfilled jobs.[24] But a trucker will quickly inform you that the industry faces a *wage* shortage, not a labor shortage—that better pay and conditions are needed to incentivize workers to take on these difficult and dangerous jobs. And indeed, it might seem economically counterintuitive that truckers' pay has been so stagnant in light of the need to attract drivers to the trade. But this is, in fact, what has occurred. As Steve Viscelli notes in *The Big Rig: Trucking and the Decline of the American Dream*, trucking firms have found that "it is more profitable to manage the [turnover] problem than to fix it."[25]

Rather than significantly raising wages or improving working conditions to improve driver retention, the industry has responded by lobbying for the removal of what they see as burdensome safety regulations, and by recruiting new demographics of workers into trucking—for example, by working to lower age limits for acquiring a commercial driver's license (or CDL) from twenty-one to eighteen.[26] Secretary of Transportation Pete Buttigieg compared these recruitment strategies to "filling a leaky bucket" in that they fail to address the root of the problem—the conditions that keep truckers from staying in the profession. A trucking advocate was more straightforward in his description of the strategy: "The reality is, if the job you're offering sucks, is the solution really to go find more suckers?"[27]

Recruitment of women has also increased in response to the shortage: women still made up only about 6.5 percent of the trucking population in 2018, but this number represents a 70 percent increase in the number of women since 2010.[28] For women and LGBT truckers, trucking can be a promising way out of rural hometowns where life is difficult or dangerous for them, as Anne Balay documents in *Semi Queer*, her excellent oral history of Black, gay, and transgender drivers.[29] But these truckers face their own hurdles in an industry that does not always welcome them: reports of sexual harassment and assault at truck stops and in driver training are legion, as are reports of gender disparities in treatment and pay.[30] Similar dynamics are at play in the racial and ethnic makeup of the industry. About two-thirds of drivers are white, and Black and Hispanic drivers comprise about 12 percent each of the trucking workforce. But the proportion of Black and Hispanic drivers attending driving schools is growing, and an influx of recent immigrants into trucking (while many of the industry's white men begin to age into retirement) is changing the demographic picture.[31]

Limited alternative employment options can keep some workers in trucking even without competitive wages. Only one in ten truckers holds a

FIGURE 2.3. The most common job in every state based on 2014 Bureau of Labor Statistics data.

two- or four-year college degree.[32] Most truckers hail from rural areas with few other prospects for employment. The map in figure 2.3—showing the most common job in every state, according to 2014 Bureau of Labor Statistics figures—reveals just how dominant truck driving is as an occupation in the states that (not coincidentally) have been hardest hit by the demise of the manufacturing labor market. Indeed, trucking is the most common job in twenty-nine US states.[33]

The turnover rate in trucking is problematic not only for truckers and firms, but for the public. New, untrained drivers are less safe than experienced ones; you'd almost certainly rather drive beside a senior driver with millions of miles under his belt than an eighteen-year-old with a fresh CDL. The public suffers financially, too—when our taxes subsidize driver training, when we pay higher insurance costs in light of more highway accidents, and when growing transportation costs raise consumer prices. In Viscelli's words, "we all pay the costs of the dysfunctional trucking labor market."[34]

The squeeze in trucking has not affected all firms equally. Trucking represents a convergence of both large and small businesses; according to 2019 Department of Transportation (DOT) data, about 90 percent of registered companies operate six or fewer trucks (and many of these are one-truck operations).[35] But by market share, trucking is dominated by a small number of very large carriers: 20 percent of trucking firms control approximately 80 percent of industry assets.

Many truckers aspire to one day own and operate their own trucks, and are persuaded to take on astronomical loans to do so. Owner-operators (or "O/Os" as they are sometimes called) compete for individual hauls, often via brokers or electronic "load boards"; these drivers make up about 10 percent of the trucking workforce. But as Viscelli documents, being an owner-operator has become a particularly risky proposition post-deregulation. Equipment prices have skyrocketed, and predatory financing arrangements often lock drivers into untenable situations. Increasingly, owner-operators work under long-term exclusive contracting arrangements with firms—meaning that the trucker bears the brunt of health and truck insurance, repairs, fuel costs, and the like, but without the ability to decide how much and for whom to haul.[36] Firms are, of course, all too happy to work with contractors rather than employees, since they do not need to pay these costs.

As operating costs in trucking rise, large companies are able to grow to accommodate them; in trucking, there are huge benefits to operating at scale across the country. Small carriers and owner-operators have a much more difficult time competing, and bankruptcies are common.[37] These two dynamics—consolidation of market share into larger firms and exclusive contracting arrangements that lead truckers to internalize risks without receiving the benefits of independence—spell danger for small business and entrepreneurship in the industry.

If the Wheel Ain't Turnin', You Ain't Earnin'

Fundamentally, truckers' overwork is a direct result of how truckers are paid for their labor. By far, the dominant mode of compensation in the industry is *pay per mile*. Most truckers are paid a flat mileage rate based on the distance between points A and B—typically without allowance for the myriad contingencies, delays, or nondriving tasks that routinely occur between A and B. This payment structure is an effort to align drivers' economic incentives with those of trucking firms, by encouraging them to maximize revenue-productive driving time and concomitantly to minimize time spent *not* moving goods around the country. In trucker parlance: "if the wheel ain't turnin', you ain't earnin'!"

But in practice, being a truck driver requires much more than driving a truck. Truckers are responsible for many different job tasks—and not all of them take place behind the wheel while the truck is rolling. Truckers must complete repairs and inspections, fuel up at truck stops (hundreds of gallons of diesel at a time), coordinate with shippers and receivers, and wait to be loaded and unloaded at terminals (a practice called "detention time" that can regularly stretch into several uncompensated hours). Most truckers make *no money at all* for the time it takes to complete nondriving tasks, or to take legally required breaks, or for time that they're waylaid by traffic, weather, road detours, or any of the other exigencies of life on the highway. Before deregulation, many truckers *did* earn hourly pay for nondriving time, in addition to mileage pay; after deregulation, the market "competed away" such pay, shifting the burden of market inefficiencies, unforeseeable delays, and mechanical risks to drivers.[38]

In most industries, the law provides a check against placing such burdens on workers. But the law has provided little recourse against truckers' economic bind. Truck drivers are explicitly exempted from the Fair Labor Standards Act, the federal statute intended to prevent labor exploitation by entitling workers to overtime pay and other protections. As we shall see, truckers' work time *is* limited by federal hours-of-service regulations—but because enforcement of these rules has traditionally been relatively weak, these rules have largely failed to provide a meaningful check on drivers' overwork.

Given the failures of law and of the market to protect truckers' interests, one might wonder whether unions might step in to do so. But collective bargaining, too, has proven a relatively ineffective instrument for protecting truckers. Unions were initially a significant force in trucking in the 1920s and 1930s; in fact, the term "teamsters" stems from the practice of "team" driving (initially horse-cart driving) in the precolonial period. Unions remained influential during the regulated era, due in part to the fact that firms lacked strong incentives to resist unions (since firms were not setting rates competitively). This was the case in spite of unions' connections to organized crime and corruption.[39] (Jimmy Hoffa, who headed up the Teamsters in the 1960s and early 1970s, was convicted of Mafia-related fraud and jury tampering, among other crimes; part of his tenure as president of the Teamsters occurred while he was incarcerated in federal prison.)

By the early 1970s, trucking was one of the most heavily unionized American industries, with over 80 percent of its workers in unions. But union power declined sharply with the advent of deregulation, as unions were forced to make concessions in light of the newly competitive nature of the industry, widespread rate and wage declines, and the flood of new market entrants. From 1975 to 2000, the rate of union membership among truck drivers plummeted from 60 percent to 25 percent,[40] and it has continued to decline since then. Though contemporary statistics are hard to come by, sociologist Steve Viscelli notes that "there are essentially no long-haul truck-load union drivers today."[41] Trucking economist Michael Belzer estimates that unions represent less than 10 percent of drivers currently on the road.[42]

My own interactions with truckers bore out a common distaste for unionization, based on the general perception that unions presented bureau-cratic limits on truckers' occupational independence, of which they are often quite protective. Perhaps unsurprisingly, truckers' attitudes toward bureau-cratic rules (both governmental and organizational) tend to be derisive. A strong libertarian streak runs through much of truckers' political rhetoric; truckers are often contemptuous of what they see as "ivory tower" hierar-chical institutions meddling in private business. As one driver put it, "[a union] doesn't appeal to me. I just like my freedom and just do what I got to do." Another told me he did not like hauling to the East Coast because he perceived a stronger union presence there; he shared a story in which he was making a delivery to a construction site and faced a confrontation over whether the work was union-protected:

> [I] got everything ready [to unload]. And the guy comes up and says this is a union job. And I . . . went ahead and continued doing what I done. And he said, you got a union card to operate that crane? I says, nope. I'm not union and neither is my truck. So he says well, I told you that it's a union job. And I says yeah, and I told you congratulations. He says well, you gotta have a union card to operate this crane. And I said, no I don't. I said I told you, unless you're deaf, I'll repeat it again, and I said, you know, me and my truck are *not union*. The stuff that I'm hauling here is not union. And it is mine 'til it touches the ground. . . . *I don't have to go by none of your rules*, you know, until it touches the ground. . . . You just run into problems like that on the East Coast a lot. And so I just, I just don't care to deal with it.

"It Takes a Special Breed to Be a Truck Drivin' Man"

As we've seen, trucking work is objectively dangerous, difficult, ill-paid, and underappreciated—it is, in many ways and unequivocally, a "bad" job. And yet, talking to truckers about their work reveals another side of the occupation. Trucking is more than a job: it's an identity, a source of pride, a tradition, a passion. This might appear counterintuitive. Yet scholars of social identity have found that people who engage in stigmatized, low-prestige jobs—what we often call "dirty work"—often *do* find positive meaning in their work and come to identify proudly with it.[43] People in "dirty" jobs sometimes develop a sense of being misunderstood by, and distant from, broader society—a belief aided by unconventional work hours or physical isolation. This leads them to develop strong occupational cultures that affirm the significance and merit of what they do (see: "if you got it, a truck brought it!").

When I spoke to truckers about what attracted them to trucking, many remarked that the independent nature of the job—the ability to make one's own day-to-day decisions about how to accomplish daily work tasks, to work hard without having a boss peering over one's shoulder, the "lone wolf" nature of the work—deeply appealed to them. Some drivers tell me they came to trucking after having conflicts with authority figures in more traditional employment settings.

Many drivers, of course, have worked in offices or in factories before, and have found that the work hasn't suited them—or that those jobs vanish from the communities in which they live—and they turn to trucking. Others get their start on farms or have served in the military. For illustration, consider the employment history of one trucker I met named Adam, a newly retired sixty-four-year-old driver with fifty years of trucking experience. Adam began driving flatbeds and combines on his family's ranch in Oregon when he was only ten years old. At twelve, he drove his first semi, his family's double-decker cattle hauler, in which he would transport cattle from the ranch to the stock pens at the nearby railhead. Adam served in the Navy, where his trucking experience secured him jobs driving buses and trucks, both in Vietnam and in Japan, where he drove a prisoner transport bus between airbases. After his military service, Adam drove freight trucks and tractor-trailers for carnivals across the western United States until his retirement.

Many truckers have professional biographies that resemble Adam's in some way. Despite its low wages, trucking has afforded many drivers a way out of even worse economic straits. Trucking skills have been essential for serving their families or their country, and have been a source of honor. This sense of honor, autonomy, and independence is thoroughly bound up with the imagery of the open road. There is romance in the way truck drivers describe the trip across the Rocky Mountains, or how the sunrise looks as they drive across the Great Plains. And many drivers talk with pride and satisfaction about how many corners of the United States they have seen

FIGURE 2.4. Trucker recruitment advertisements play upon the necessity of providing for one's family and the desire to escape a traditional "boss."

FIGURE 2.4 (*continued*)

in the course of their work—views far more majestic than the drab walls of a cubicle or the factory floor. Trucking work is tedious, grueling, and dangerous work—but it brings with it an unmatched romance and a fierce sense of pride.

The cultural iconography of the trucker evidences these values and mythologies. The trucker of film and song—nearly always a man—is depicted as strong, rugged, and robust; the truck itself operates almost as a mechanical extension of his masculinity.[44] Driving a truck is hard work that takes control, confidence, and a certain form of grace, which feed back into the way truckers see themselves. As Anne Balay and Mona Shattell, two scholars of trucker culture and health, put it: "Behind the wheel of a semi, drivers feel connected to—almost one with—all that power."[45]

In trucker culture, the trucker is portrayed as being unafraid of danger, sexually virile, unfettered by social ties, and—importantly for our story—wily enough to outsmart and overcome the misguided and bungling efforts of law enforcement officers or bureaucratic rules. In addition to visual representations, country music has historically played a role in forming truckers' cultural identity, even as it creates stylized mythologies of what day-to-day trucking life is actually like. As Merle Haggard sang in "Movin' On," his 1975 trucker anthem: "It takes a special breed / to be a truck drivin' man." Johnny Cash echoed the same sentiment in "All I Do Is Drive," in which his trucker narrator proclaims that "I got nothin' in common with any man / who's home every day at five!" Historian Shane Hamilton explains:

> [M]usicians who recorded early trucking songs adapted country music's long-lived theme of the "wanderer." By singing about stopping at a "roadhouse in Texas / A little place called Hamburger Dan's," Terry Fell helped promote the idea that a trucker's life consisted mainly of playing pinball while sipping coffee poured by a red-headed waitress. The truck driver, as a man in control of his time and possessing an untamed sexuality, was the antithesis of the "organization man" . . . that sociologists of the 1950s identified as the primary victim of an increasingly corporatized, bureaucratic urban society. Few truckers would have taken Terry Fell's song as an accurate representation of their work culture, yet few would dispute that the sense of mobility that trucking provided was an important attraction of the work. When a man's workplace was the road, a trucker took no orders from the factory foreman and faced no line speedups or stopwatch-toting scientific managers.[46]

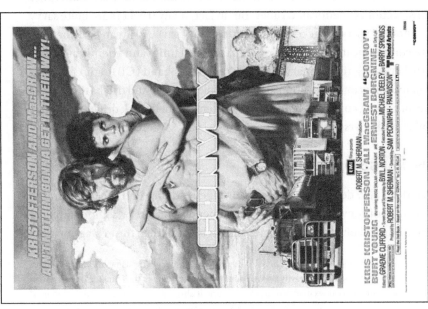

FIGURE 2.5. The cultural iconography of the trucker as expressed in film.

Truckers' emphasis on freedom and independence correlates with the gendered nature of the trucking workforce: about 93 percent of drivers are men.* Masculinity is a highly valued trait that manifests itself in a number of ways. Some truckers I met were self-described "family men," with strongly professed religious values and traditional notions of gender roles, who proudly showed me pictures of children and grandchildren they were helping to provide for through their labor.[47] For others, masculinity expresses itself as machismo—some truckers told me boastfully about their varied sexual exploits as their work took them across the country. This "cowboy" mentality is evidenced in the genre of trucker films, such as *Smokey and the Bandit, Convoy, Over the Top*, and *White Line Fever*, which glorify disregard for (usually corrupt or inept) authority figures, and which feature strong, fearless truckers as leading men. No matter how it is manifested behaviorally, being "manly" seems to be a particular point of pride for many truckers I met, bound up closely with the independence of the job—and, as we shall see, a particularly salient driver of their relationship with technology.

Truckers do not fit the stereotype of a technology consumer or early adopter. Their rural backgrounds, frequent lack of formal educational credentials, and blue-collar work engender relatively little attention from technology consumer markets; when they *are* considered, they are sometimes constructed as uneducated rubes. But in fact, many truckers are quite savvy technology users. Truckers' constant mobility gave rise to an early need to use communications technologies to talk to family and friends at home. Truckers were the primary adopters of citizens band (CB) radio for communication prior to the popularity of mobile phones. And trucks themselves are, of course, complex technological systems, which truck drivers routinely tinker with, personalize, and repair.

The Not-So-Open Road

Paradoxically, despite their desire for independence and escape from the strictures of conventional work environments, truckers encounter far *more* day-to-day involvement with rules and regulations than the average person.

---

* Readers may notice that in this book, I frequently refer to truckers using he/him pronouns. I do this in recognition of the heavily skewed demographic makeup of the trucking population as well as the centrality of masculinity to trucker culture. It also reflects the self-identification used most frequently by the truckers with whom I spoke when conducting my research (which I discuss in more detail in Appendix A). I highly encourage readers interested in the unique experiences of women, LGBT, and nonbinary truckers to consult Anne Balay's oral history about these drivers, *Semi Queer.*

Some of these entanglements are based on firms' organizational rules and employment or leasing arrangements. Others stem from the fact that truckers' daily work takes place largely on federal highways; it is on the system of public roads—with their speed limits and traffic controls, insurance and licensure requirements, laws about seatbelt and cell phone use, and the like—that most of us feel the presence of behavioral rules most directly. Add to this the additional highway rules that apply specifically to truckers (designated "no-truck" routes, weight and height clearance limitations, fuel tax regimes) and the dozens of details of the trucker's workday regulated by the federal government, and the "open road" begins to appear much less open. As Hamilton characterizes it, today's truckers must contend with a "dense web of weigh stations, ports of entry, reams of paperwork, layers of taxation, and contradictory regulations."[48] What's more, economic deregulation has compelled an *increase* in "social regulation"—federal rules, structures, and bureaucracies that govern firms' practices and truckers' working conditions within the newly competitive market, especially around issues like driver overwork and fatigue.[49] Ironically, as Ronnie Milsap sings in his famous trucker anthem, truckers find themselves "imprisoned by the freedom of the road."

A myriad of controls on truckers' day-to-day activities come from the Federal Motor Carrier Safety Administration (FMCSA), the federal administrative body—a part of the US Department of Transportation—responsible for issuing and enforcing regulations involving commercial motor vehicles. In the name of furthering the agency's safety mission, the FMCSA brings dozens of details of a truck driver's workday under the purview of federal regulations. The Federal Motor Carrier Safety Regulations[50] are the body of rules governing every aspect of commercial vehicles' operation on American highways, from safety requirements for drivers and equipment to what trucks may haul, from procedures for inspection to rules for noise and emissions. The regulations are extremely detailed. The "Green Book," which compiles them in a single volume to be easily referenced by drivers, runs to about six hundred pages of fine print—what one trucker termed "a book with the thickness of the Bible."[51] These rules range from strict training and licensure requirements and screenings, to restrictions on the hours a trucker can drive and the paperwork he must keep as evidence of his compliance, to the thorough vehicle inspection he must perform and document each day. There are rules about how much noise a driver's truck can make, what type of fuel he may use to fill it, and what he may transport in it. Some of the rules pertain intimately to the driver's body. Truckers must submit to drug and alcohol testing and must pass mandatory licensure physicals ensuring they are in good enough health to drive.

But by far, the most economically important and highly regulated aspect of a trucker's life is *how much time he is allowed to drive*. It's this question to which I turn next.

3

Tired Truckers and The Rise of Electronic Surveillance

Perhaps no aspect of long-haul trucking work is more important, and more controversial, than sleep. When and how much to sleep (and concomitantly: when, how much, and under what conditions to drive) is a key logistical consideration in trucking, and one with crucial economic, safety, and cultural implications. And it's the backdrop against which we'll consider how surveillance technology has entered the truck cab, and what that means for truckers and their work.

It's abundantly clear that truckers are overworked and underslept. For reasons we'll discuss, it's difficult to attach a number to exactly how many hours per day or per week truckers actually work. One pre-ELD study by the University of Michigan Transportation Industry Program found that non-unionized long-haul drivers worked sixty-five hours per week; 10 percent of respondents in that study reported working *more than ninety-five hours* in the previous week.[1] In another survey, a third of long-haul drivers admitted to falling asleep at the wheel at least once in the previous month.[2] Sleep studies conducted among truckers consistently show that truckers average only five to six hours of sleep in a twenty-four-hour period when they are working.[3] Short sleep duration and low sleep quality among drivers can lead to diminished attention, memory, and cognitive response time while driving.

Trucker fatigue can have deadly consequences, contributing to thousands of fatalities and billions of dollars in losses every year. Each year, truck crashes on America's highways kill about five thousand people and injure 150,000 more—figures that have been climbing steadily over the last decade.[4] Though several factors contribute to crash risk, driver fatigue has been repeatedly identified as having one of the strongest effects on accident rates, particularly in fatal crashes.[5] For instance, in one emblematic case, federal regulators shut down a trucking company after one of its sleepy drivers killed an Illinois Tollway employee and injured a state trooper while they were assisting another driver at roadside. A subsequent investigation revealed that during the twenty-six hours immediately preceding the crash, the driver had driven about one thousand miles and had rested for only between three and six hours.[6] In 2014, the problem of tired trucking drew some (brief) public attention after a Walmart trucker crashed into Tracy Morgan's tour bus, killing comedian James McNair and seriously injuring Morgan and four others. The truck driver was alleged to have worked without sleep for more than twenty-four hours straight before the crash.[7]

Why are truckers so tired? Structurally, the biggest reason is the way truckers are paid. As I discussed in chapter 2, the pay-per-mile arrangement that is standard in the industry incentivizes truckers to overwork. Remember: if the wheel ain't turnin', you ain't earnin'! Time spent sleeping is time spent not driving—and that means less money in the trucker's pocket.

These economic incentives to overwork are mirrored in trucker culture. The idea that "sleep is for sissies," as described by historian Alan Derickson in his book *Dangerously Sleepy: Overworked Americans and the Cult of Manly Wakefulness*,[8] finds expression in occupations as diverse as investment banking, software engineering, factory work, aviation—and truck driving. As Derickson tells it, deep American historical roots connect Protestant notions of hard work, manliness, virtue, stamina, and success with the denial of sleep. Working while tired can be construed as "evidence of dedication and personal strength."[9] In his 1994 book *Pedal to the Metal*, Lawrence Ouellet describes how truckers reproach one another for not being man enough to "stay in the saddle,"[10] and observes that "[truckers] displayed a certain pride in endurance that was reflected by a slang term peculiar to them, 'doing the double.' This term denoted working two consecutive shifts[.] . . . [T]he driver who couldn't stay in the saddle . . . was seen as lacking skill (endurance)."[11]

The lack of sleep, and the lengths to which truckers will go to stay awake, have long been glorified in truckers' "war stories" and in media. Consider Dave Dudley's famous confession in "Six Days on the Road" that "I'm takin' little white pills / My eyes are open wide" (referring to amphetamines). As Derickson explains, truckers throughout history have reportedly relied on all manner of tricks to stay awake; a 1935 National Safety Council research paper he cites reported that "[a]mmonia as an inhalant is useful, but its effect is not lasting. Some drivers are even reported to resort to sipping urine in emergencies."[12] Other truckers "us[ed] an onion to moisten dry eyelids."[13] Others, afraid of nodding off while driving, "would light a cigarette and sleep until it burned down and awakened them by scorching their fingers"— this particular strategy was described by Jimmy Hoffa in 1936, marshaling trucker fatigue as a motivation for enrolling truckers in collective bargaining arrangements.[14]

Stimulant use by truckers remained a known public health problem throughout the twentieth century, and was often cited by government agencies as a rationale for antidrug regulations. Truckers had both the means and motivation to obtain chemical aids to stay awake, due to the endurance needed for their work and their access to (and sometimes participation in) drug trafficking networks over the federal highways.[15] Truckers sometimes referred to letting "Benny take the wheel"—a reference to Benzedrine, a form of amphetamine; the knowing reference was even made in popular trucker songs (like Commander Cody's 1972 anthem "I Took Three Bennies and My Truck Won't Start"). In the 1980s, the Drug Enforcement Administration banned Ecstasy (a combination of methamphetamine and mescaline) after truckers began using it to stay awake for work.[16]

Randomized drug testing in the industry makes drug use a somewhat riskier solution to fatigue in the modern era, though several truckers told me that stimulants can still be purchased under the counter at many major truck stops. The use of legal stimulants like 5-Hour Energy and Red Bull is extremely widespread. Lack of sleep is viewed as bravado and a sign of work ethic. In 2015, Florida Highway Patrol officers investigated a driver after he bragged on Facebook about driving twenty hours without sleep, saying: "Lakeland Florida is 1,470 miles away it gonna take 20 hrs to do this run [. . .] forget sleeping $$$$ talk."[17] (The driver claimed he was merely joking with friends.) Other drivers report keeping a bowl of ice water next to them to splash on their faces when falling asleep.[18] Companies often know about, and even encourage, these practices. A Texas trucking company recently faced flak for a meme it posted to its Facebook page that read: "DON'T STOP WHEN YOU ARE TIRED. STOP WHEN YOU ARE DONE!"[19]

It's important to recognize statements like these for what they are, and not to caricature or oversimplify them. Truckers' performance of masculinity as stamina, along with the cultural iconography and "cowboy" identity that attend the profession, can be understood in part as cultural glosses necessitated by trucking's political economy. As we've discussed, truckers take part in the exploitation of their own labor by "self-sweating"—collaborating in the process of exploiting their own labor[20]—and by resisting or circumventing safety regulations based on economic necessity. Much of the machismo, and the "fun and games" of outsmarting law enforcement, as depicted in truckers' casual discussions and in cultural iconography, is posturing that masks the tight economic straits in which they find themselves. Oral historian Anne Balay describes this "trucker mentality" as "confidence born of years of experience mixed with sheer desperation."[21]

One trucker I spoke with described this far more powerfully than I ever could. As I was wrapping up our interview, he stopped me to interject:

> It's just for me, when you talk about breaking hours of service, and you talk about running multiple logbooks, and you talk about, you know, electronic regulation and things like that, . . . [it's] important for me to touch upon the financial conditions which produce, like, that kind of behavior[.] I mean, it's not a tenable situation without some kind of help. You know, it's not something that you could do. And there are a lot of men out there who—there wouldn't be food on the table, frankly, and the lights wouldn't be on at home if they weren't breaking the law and if they weren't using drugs. And it's not about having a party, because it's not a fucking party. It's very much not a fucking party. I mean, you're driving something that weighs from about 65,000 to maybe 100,000 pounds. You can go into a ditch. You could crash into somebody. You could rear-end someone. And you've been awake for 20 to 30 hours. There's nothing fun about that. . . . You're looking in the rear-view mirror. You're looking in the ditch for a smokey. You're worried about the chicken coops being open. It's not fun. That's not fun.

Hours and Logs

In an attempt to address the fatigue problem, the federal government has legally limited truckers' work time through "hours-of-service" regulations. These rules have been in place, in one form or another, since the 1930s.

Though specific time limits have changed over the years, most long-haul truckers today may legally drive no more than eleven hours per day, and can be "on duty"—a status which includes both driving and other job functions, like fueling, loading/unloading, and performing required vehicle inspections—for no more than one fourteen-hour window per day. After the maximum on-duty time is reached, a ten-hour break is required. Weekly limits also apply: most truckers can work up to seventy hours in any eight consecutive days, and then must break for thirty-four hours (known as "resetting your 70").[22] The hours-of-service rules apply specifically to long-haul operations—in which trucks drive "over the road" for long distances, typically interstate—in which fatigue is of greatest concern. (Short-haul operations, which are legally restricted to operate within a 150-mile radius of the truck's home base, and delivery services are subject to different regulations; my exploration in this book focuses on the long-haul segment of the industry specifically.)

To facilitate enforcement of the hours-of-service regulations, truck drivers have long been required to keep track of their work time using logbooks. The daily log was initially mandated by the Interstate Commerce Commission (ICC)—the predecessor to today's Federal Motor Carrier Safety Administration, or FMCSA—in 1940. For the last forty years, these records have been kept using a standardized "graph-grid" form.[23] Graph-grid paper logbooks have long been a staple for truck drivers; a thirty-one-day supply can be purchased at any truck stop for a couple of dollars. Logs must be kept up to date at all times, as they can be audited by law enforcement at weigh stations or at roadside—but importantly, these compliance documents are typically *not* used to compute drivers' pay.

The graph-grid log, pictured in figure 3.1, contains lines for a driver to record his duty status, a section for a driver's remarks, and other general information about the carrier and driver. The ICC's initial regulations required that the log "bring out the essential facts" about where the driver had operated within a twenty-four-hour period; how long the driver had been on duty; and how much of this on-duty time was spent driving. The driver was required to keep a copy of the log with him and to file a copy with his carrier each day or upon returning from a trip; carriers were required to retain these logs for inspection. Though the form of the log was occasionally tweaked over time, it remained, in its essence, a paper-based record of a driver's daily activity.

The graph grid is easy to understand and to fill out. It is comprised of four rows, representing the four possible duty statuses a trucker may have: he may be off duty; he may be at rest in the truck's sleeper berth (similar

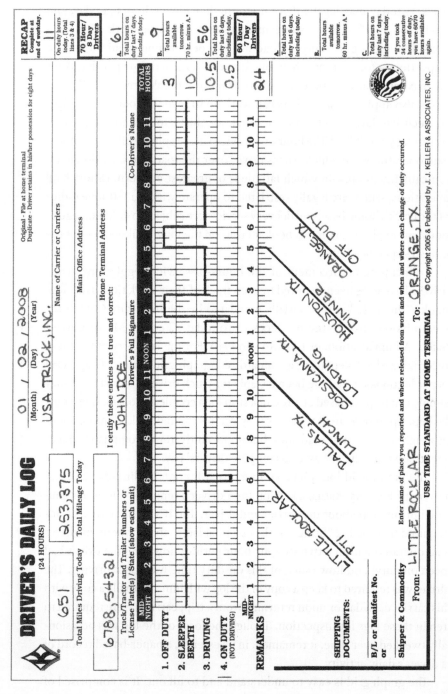

FIGURE 3.1. Paper logbook entry page. The driver records duty status by drawing a horizontal line on the appropriate row of the grid and manually calculates the hours spent in each status at the end of the row. Illustration by author.

to being off duty, but a status to which some additional timekeeping rules apply[24]); he may be driving; or he may be on duty, but not driving (conducting inspections, being loaded or unloaded, or completing other essential tasks). Columns correspond to the hours of the day. The driver records a duty status by drawing a continuous line across the grid in the appropriate row to represent the period of time he spent in that duty status. When his status changes, he "jogs" the line to the row for his new status and records the name of the city or highway mile-marker where the change took place. At the end of the day, the driver is required to tally the number of hours he spent in each duty status and compute the number of hours, if any, he has available for work for the following day. Drivers are required to keep the previous seven days' worth of logs available for inspection and must file their logs with employers within thirteen days; employers must retain the logs for six months.[25]

Swindle Sheets

For as long as the timekeeping regulations have existed, truckers have fashioned ways to evade them. As we have seen, drivers face strong economic pressures to maximize their driving time and are incentivized to remain on the road even when exhausted. What's more, some employers impose deadlines that cannot be met if the rules are strictly observed, and independent drivers who negotiate with brokers for each haul may altogether miss opportunities for work if they toe the line too closely. Trucking firms may explicitly or implicitly pressure drivers to underreport their hours in order to move goods at the pace of work demanded by the market. Even when a trucker *wants* to stop for a break, the dearth of truck parking along American highways—thanks in large part to local NIMBYism based on truck stops' unglamorous reputations—can make doing so infeasible.

Thus, it is widely acknowledged that truckers lie on their logbooks. They do so routinely—so much so that they are often dismissively referred to as "comic books," "coloring books," or "swindle sheets." Recall that these documents are used for regulatory and compliance purposes, but typically not for driver compensation; thus, adjusting a log doesn't have a detrimental financial effect for a driver, and much more likely frees him up to make more money. In one survey, only 16 percent of drivers reported believing that logbooks provide an accurate depiction of drivers' activities;[26] another study found that nearly three-quarters of all drivers admitted to violating the hours-of-service regulations.[27] The FMCSA has consistently described widespread hours-of-service violations as one of the major problems plaguing the industry.

Longstanding references in trucker popular culture suggest the ubiquity of these violations and the extent to which they are (not) taken seriously by truckers. Consider the 1963 trucker anthem "Six Days on the Road," written by Carl Montgomery and Earl Green and popularized by Dave Dudley ("I.C.C. is checking on down the line / I'm a little over weight, my logbook's way behind / But nothing bothers me tonight / I can dodge all the scales all right / Six days on the road and I'm gonna make it home tonight") or the 1975 hit "Convoy" by C. W. McCall ("We tore up all of our swindle sheets / and left 'em settin' on the scales").

Log-fudging is technically simple. Though logs are required to be kept up to date, they can easily be adjusted via "creative writing" at the end of the day to make a trip look like it was "run legal" even when the driver exceeded his maximum time. Law enforcement officers at weigh stations can be avoided with relative ease, particularly when drivers exchange information among themselves as to whether a station is open or closed—in CB slang, whether "the coop is clean"—and plan their routes accordingly. (The "coop"—short for "chicken coop"—is common trucker parlance for an inspection station or scale house where law enforcement officers are; when "the coop is clean," the weigh station is not occupied by law enforcement and can be bypassed. The phrase derives from another well-known term in the trucker vernacular: referring to law enforcement officers as "chicken shit.")

The Digital Tattletale

Tired truckers, undeterred by easy-to-flout timekeeping rules, evidence a widespread and extremely dangerous problem in search of a solution. The problem's roots are economic—to make a living, truckers have little choice but to break the law and to put themselves, and the motoring public, in danger. One might imagine that this problem has an economic solution, rooted in changing the structure of compensation in the industry to prevent truckers from being put in such dire straits.

Yet the solution on which hopes have been pinned is not economic, but technological. Rather than change the underlying conditions that give rise to lawbreaking, regulators and companies have tried to make it more difficult for truckers to falsify their time records. They have done this by mandating the use of digital devices, integrated into trucks themselves, that create a record of the hours the truck is driven (as well as, often, other information about the truck and its driver's behavior, as I will describe in chapter 4). These devices—known as "electronic logging devices" or ELDs—automate many of the functions served by paper logs in an effort to curtail unsafe practices. The core technology of the ELD is simple; it is a small device hardwired to the truck's engine that captures data about (at a minimum) the truck's engine status, location, and mileage. Those data are tracked and displayed on an in-cab interface or, in more recent models, transferred wirelessly to a separate device like a tablet or phone—and often to a company's central dispatch office as well, so that the truck can be monitored remotely. (I'll discuss what these offices do with ELD data in the next chapter.)

The road to mandatory ELDs was long and rocky. The devices made their first appearance in federal regulations back in 1988, when the regulator permitted *voluntary* use of automated devices for compliance with the hours-of-service rules. A mandate for all trucks was proposed in 2003, but ultimately not adopted, for reasons including significant doubts about the efficacy and costs of such a rule. A 2010 rule would have mandated ELD use for "pattern violators" of the timekeeping rules: carriers with high rates of noncompliance with hours-of-service regulations, who were assumed to be among the most rampant falsifiers of paper logs. The passage of this rule was delayed when the Owner-Operators Independent Drivers Association, or OOIDA, challenged the process by which this rule was enacted. OOIDA argued that FMCSA hadn't seriously considered the prospect of firms harassing drivers via electronic monitoring—a concern that had arisen in public comments during the rulemaking process, and which FMCSA was legally required to weigh in its decision-making. The Seventh Circuit Court of Appeals agreed and vacated the rule.[28]

In July 2012, President Obama signed into law the MAP-21 highway bill (Moving Ahead for Progress in the 21st Century Act). Among its provisions, MAP-21 directed FMCSA to issue a rule requiring electronic logging in the trucking industry. The MAP-21 bill was the first Congressional mandate to *require* such agency action, making some form of ELD mandate essentially a foregone conclusion. The FMCSA's final rule mandating ELDs for nearly all American trucks running long hauls was finally issued in 2015, and the mandate took effect in December 2017.[29]

Over the nearly three decades in which electronic monitoring was under consideration in one form or another by regulators, the technology generated vigorous, often quite heated disagreement within the trucking industry. Though large trucking firms expressed some initial hesitation regarding

ELDs for a variety of reasons (including questions about accuracy and tam-perability), they eventually came to strongly support the ELD mandate—as did insurance groups and public safety coalitions. For reasons I'll discuss in detail the next chapter, large companies had both less to lose and much more to gain from the mandate. By the mid-2010s, many of them were already using electronic monitoring to manage their drivers, so the mandate didn't change their practices very much—and the collateral management benefits of the device outweighed their concerns about lost productivity. One indus-try survey conducted just prior to the mandate (in 2016) estimated that about 80 percent of large carriers had already installed ELDs, while only about 30 percent of small carriers had.[30]

But the mandate was ardently opposed by many (though not all) drivers and small companies, particularly independent owner-operators. Most small carriers were not already using ELDs before the mandate because they had very little to gain from doing so, and it was more difficult for them to bear the costs of the devices and service plans—costs that all were to be borne by truck owners themselves under the new regulations. Small carriers implored FMCSA to take their size into account when evaluating the costs and benefits of promulgating the rule, and potentially to exempt them from the mandate based on their size; FMCSA refused, opining that it was "not required or expected to provide an exception to its safety rules based solely on the fact that the businesses are small."[31] Costs of ELDs have varied widely over the years; at the time of the mandate, devices cost somewhere on the order of seven hundred dollars[32] (plus, in some cases, service plans and maintenance costs).[33] But these figures don't account for lost productivity due to sharply reduced flexibility, nor for disparate management gains to be realized by large carriers who could deploy the technology at scale.[34] (As we will see in chapter 6, a small crop of "dumb" ELDs targeted to small carriers—devices with purposely limited capabilities, some of which were purposely designed to make it difficult for law enforcement to access their data!—emerged over time. These tended to be less expensive, on the order of a couple hundred dollars, and did not require a service plan.)

In fact, some supposed that large carriers backed the ELD mandate *because* of the negative impact it would have on small businesses—because the requirement disproportionately affected small carriers negatively, it could serve to prevent them from competing with large firms.[35] Industry groups representing large firms not only supported the mandate, but even implored state legislatures to follow the feds' lead and mandate ELDs for *intra*state trucking.[36]

As you may surmise based on our discussion of trucker economics and culture, most truckers (and independent owner-operators in particular) were vocal opponents of the ELD mandate at all stages of the game. A paramount concern among truckers was economic. ELDs imposed additional costs (hardware, service plans) on already cash-strapped drivers, and many truckers worried that the mandate would reduce their ability to make money by placing inflexible constraints on their driving time. Some owner-operators also reported that the rigidity of electronic timekeeping induced them to make shorter (and somewhat more predictable) hauls, so they could avoid the risk of maxing out on hours and being forced to take a thirty-four-hour break—even though shorter hauls mean more time spent unpaid dealing with pickups and deliveries. At the same time, bigger companies with larger driver pools viewed longer hauls as more desirable, because they could take advantage of relay systems or team operations to avoid running into hours-of-service problems.[37]

Even concerns like finding parking—which might seem like a minor annoyance—could be significantly more costly in time (and therefore money) under an inflexible monitoring scheme. Truck parking is already a scarce resource; an American Transportation Research Institute study before the mandate found that for drivers being monitored digitally, parking needs turned from a headache into a real safety issue.[38] A driver using paper logs, faced with a full truck stop, might hunt for parking at the next stop even while officially off duty. But because the ELD automatically detects a truck's movement, doing this might thrust a digitally monitored driver from being off duty to being on the clock again—potentially putting him in violation of the rules. A digitally monitored trucker thus has much less flexibility about when he can stop for the night; as one driver lamented, "ELD leaves no room for dealing with crowded truck stops, making it nearly impossible to preplan."[39] To avoid this, monitored truckers may simply pull over on a highway shoulder or off-ramp to take their ten-hour break—an unsafe (and often illegal) practice that endangers other drivers.

Even more fundamentally, truckers raised concerns about ELDs as an affront to their privacy, dignity, and independence. A common refrain was that the new rules treated them like criminals, cheaters, or children: as one put it in comments sent to FMCSA, "I do not need a federal baby-sitter in the truck." Another: "I resent the need for a black box on my truck to track what I am doing. I am a safe and honest driver."

Truckers brought (ultimately unsuccessful) legal arguments to challenge the ELD mandate on privacy grounds. OOIDA, the owner-operators'

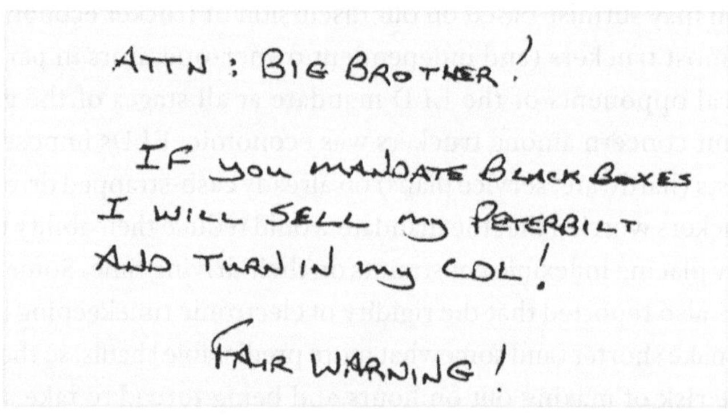

FIGURE 3.2. A public comment submitted to the Federal Motor Carrier Safety Administration while the agency considered mandating ELDs. Peterbilt is a well-loved truck manufacturer; a CDL is a commercial driver's license, which truck drivers are legally required to obtain.

association, sued FMCSA again in 2016 to attempt to block the mandate—asserting, among other arguments, that the rule violated the Fourth Amendment's prohibition on warrantless search and seizure.[40] The Seventh Circuit Court of Appeals rejected this argument under the "pervasively regulated industries" exception, which holds that businesses that operate voluntarily in highly regulated industries (like liquor stores and gun dealers) can be required to undergo administrative inspections without implicating the Fourth Amendment, particularly if they engage in dangerous activities. Regulatory inspections are seen as part of the bargain for doing dangerous business. Trucking is without question both a dangerous and highly regulated industry, as we have seen, thus precluding truckers' legal basis for challenging the rule on privacy grounds.

But it's understandable why this rationale carried little weight for truckers. The truck cab is not merely a workplace—it's a trucker's home. It's an intimate space, where he sits for hours alone. He sleeps and eats there. It is his small corner of the world, from which he can exercise some control. In those ways, it's wholly unlike a liquor store or a gun shop—electronic monitoring can feel like an infiltration into a much more sacred space.

A related point of pushback from truckers concerned their prized autonomy—to make decisions, to do their work as they see fit, and to know their own physical capabilities and limitations. As we have discussed, truckers often attach tremendous value to the independence of their work. The hours-of-service rules, as well as the electronic monitoring used to enforce them, challenge this independence, and (on one reading) imply that not only does the trucker lack the stamina needed to do the job, but that he's too weak and faulty to *know* when he's too tired to drive. (In chapter 4, I'll describe further how the ELD seems to deprioritize certain forms of knowledge, including biophysical knowledge, traditionally held by the trucker alone.) This sentiment is well-captured in the following comment, submitted by a trucker to the FMCSA in 2011:

> A computer does not know when we are tired, [f]atigued, or anything else. Any piece of electronics that is not directly hooked up to my body cannot tell me this. . . . I am also a professional [and] I do not need an [ELD] telling me when to stop driving . . . I am also a grown man and have been on my own for many many years making responsible decisions!

This comment, as we shall see, is accidentally prescient—in positing that a computer "not directly hooked up to his body" can't detect his level of fatigue, he certainly does *not* seem to be suggesting that the ELD *should* be directly hooked up to his body! Indeed, in the Seventh Circuit's 2016 ruling upholding the ELD mandate against the challenge from OOIDA, it speculates on hypothetical alternatives to the tracking enabled by the ELD—"constant video surveillance or perhaps some form of bio-monitoring device"—only to dismiss these alternatives as "breathtakingly invasive" and therefore not what Congress must have envisioned when it mandated automatic tracking.[41] As we shall see in chapter 7, the use of either of these might not have been as far-fetched as either this trucker or the Seventh Circuit might have envisioned.

Truckers are, of course, a large and diverse group of workers—and within any group of nearly two million workers, there are bound to be differences of opinion on any issue. A minority of truckers did speak in favor of ELDs. Some of these truckers spoke of "leveling the playing field." Some drivers said that they felt undue pressure to break the timekeeping rules; despite their reluctance, they felt they had to skirt the rules because they were competing for loads against drivers who cheated wantonly. These drivers felt that ELDs would prevent these flagrant rulebreakers and allow them to "run legal" with less competitive pressure. Some drivers also hoped that ELDs would give them a defense against dispatchers who might pressure them to "make a load work" even if doing so would require breaking the law.[42]

Missing the Forest for the Trees

Another key concern that truckers brought to the fore was that the ELD addressed the wrong problem. Many truckers agreed that they were under time pressure in their work, and that they often didn't get enough rest to drive safely. But focusing on truckers' log fudging treated a *symptom* of the problem, not its root cause. Truckers weren't tired *because* they were able to falsify their logbooks; they were tired because the industry was set up in ways that necessitated them breaking the rules. By focusing on what *truckers* were doing to react to that system, the ELD mandate viewed them—the least powerful members of the industry—as untrustworthy liars who needed to be better policed, rather than professionals doing their best to negotiate difficult logistics in the face of countervailing demands.

And what *is* broken about the system? The pay-per-mile arrangement dominant in the industry is a big part of it. Pay-per-mile is a form of piecework: it rewards truckers for what they "produce" (here, how far they move their haul) but *without* accounting for all the other work it takes to make that movement possible. That work—time for inspections, for loading and unloading, and indeed, for rest—is typically unpaid and unacknowledged.[43]

What's more, it's that unacknowledged, unaccounted-for work time that, truckers say, is the *real* driver of fatigue in the industry! Truckers pointed overwhelmingly to a chief example: what is known in the industry as "detention time."[44] When truckers arrive at a shipper's or receiver's terminal, they don't immediately drive up to be loaded or unloaded. Instead, they often face *very* long (and unpredictable) delays while they wait for a dock and personnel to become available. There are a number of reasons for this—personnel might not be incentivized to work quickly, unpredictable arrival times can impact receivers' ability to load or unload in a timely way, shippers and receivers may overbook appointments. In one survey, 63 percent of truckers reported typically waiting more than three hours at a shipper's dock.[45] Another study by Virginia Tech's Transportation Institute found that drivers spent, on average, 3.4 hours waiting at a customer's facility.[46] Women drivers report more delays than men—which some chalked up to the fact that men were more likely to "express more consternation" to dock personnel when they had long waits.[47]

This time is *generally unpaid* for truckers: they are paid by the mile, and they are barely moving during this period. A 2018 DOT study estimated that trucker detention cost drivers and carriers between $1.1 and $1.3 billion in earnings each year.[48] But though this time is unpaid, it is still work; a trucker can't merely head into the sleeper berth and wait to be unloaded. He is still responsible for making "yard moves" at the terminal, for keeping tabs on his freight, for communicating with terminal personnel, and other tasks. Even when some trucking companies have requested that they be able to count that work time as off duty—thereby preventing them from wasting all their driving time waiting around—FMCSA has refused, noting that such a change would risk creating "the potential for extremely long work days."[49]

Detention time is a problem because there isn't much incentive for shippers to be efficient once a truck arrives at the terminal—but it's truckers who pay for this inefficiency. Some shipping contracts do account for detention time, charging an hourly fee to shippers if drivers are detained (usually construed as more than two hours' wait)—but not all of that money accrues to drivers.[50] Further, it can be more difficult for small companies and independent drivers to include these clauses and remain competitive with larger companies.[51]

Unsurprisingly, detention time causes significant problems for hours-of-service compliance, as a driver facing a long delay may run out of available hours to drive to his next destination. Carriers classify facility delays as the single biggest factor impacting drivers' ability to comply with hours-of-service rules.[52] Eighty percent of drivers report that excessive detention time had made them late to a subsequent job or had led them to cancel a scheduled pick-up in the prior year.[53] And detention time is directly traceable to safety outcomes. A DOT study estimated that only fifteen minutes of detention time (beyond the two hours' wait that is deemed reasonable to complete a loading or unloading) can increase crash risk by 6.2 percent—which translates to 6,500 extra truck accidents per year.[54] What's more, every 5 percent increase in the number of stops that lead to detention time results in a 4.7 percent increase in expected accidents.[55]

These statistics tell a clear story. Detention time is dangerous, costly, and has no obvious benefit. If regulators want to reduce fatigue and prevent truckers being forced to "race the clock," finding ways to reduce detention time is essential. But FMCSA has only recently begun to explore what steps it might take to address detention time, under pressure from lawmakers.[56]

Does Electronic Monitoring Solve the Safety Problem?

It's worth asking, of course, how well ELDs address the problem to which they are supposed to be the solution. We might reasonably feel that the downsides of trucker surveillance are justifiable if they meaningfully address the high-stakes problem of danger on the public roads. Does electronic monitoring prevent truckers from breaking the hours-of-service rules? And perhaps even more to the point—do they make our public highways safer? It's important to recognize that these policy goals, while related, are not quite the same; the technology may lead to more measurable compliance with the regulations—but that doesn't necessarily lead to the actual safety outcomes that the regulations are designed to operationalize.

In fact, the verdict on these questions is decidedly mixed. The most rigorous study of the ELD rollout, conducted by three business school professors,[57] found that the mandate *did* increase drivers' compliance with hours-of-service regulations, as measured by the number of citations issued by law enforcement during inspections. The change was driven primarily by small trucking companies' practices, because large carriers—seeing the writing on the wall, and having their own reasons for deploying the technology, as I'll discuss in the next chapter—had already adopted ELDs by the time the mandate was finalized. Before the mandate, inspectors cited drivers for hours-of-service violations about 6 percent of the time; afterward, they found (or at least, recorded) violations only half as often, 2.9 percent of the time. The biggest reduction was for small carriers. (More on these numbers, and why they may not mean exactly what they appear to, in chapter 5.)

Among other things, the ELD mandate virtually did away with "form-and-manner" citations—tickets issued for not filling out one's paper logbook properly—which had been the most frequent form of timekeeping violation.[58] This sort of violation sometimes reflected a driver's intentional falsification of his log—but sometimes just reflected a driver who'd made a mistake on his arithmetic or had otherwise been sloppy on his recordkeeping (without necessarily reflecting that he had also been driving too much).

But crucially, the study did not find that the rollout of electronic monitoring led to *any improvement* in the safety outcomes we actually care about! Truck crashes didn't decrease after the mandate began to be enforced—and for small carriers, they actually *increased*.[59] The number of fatalities in large-truck crashes hit a *thirty-year high* in the first year of the ELD mandate, even as general vehicles fatalities decreased.[60]

The reasons for this are not difficult to fathom, even if you're not a trucker. Imagine you are told you have *about* ten hours to complete a six-hundred-mile trip to see your grandmother. But you know that if it actually takes you ten-and-a-half hours or even eleven hours, no great negative consequence will befall you; your grandmother will just have to wait a little while. If you drive at a steady sixty-mile-per-hour rate, you should just make your projected arrival time. And because you can sometimes drive a bit faster—say sixty-five or seventy miles per hour—during your trip, you have time to take bathroom and stretch breaks, to grab a snack, to slow down in inclement conditions, to get gas, to take a quick look at something that doesn't sound quite right on your vehicle. It's certainly still hard work, and you can't dither around—but you *can* do the things that allow you and your vehicle to arrive safely and in good health at grandma's house in ten-ish hours. Indeed, this is the way most of us think about travel estimates on the road. It's nearly impossible to predict to the minute exactly when you'll arrive somewhere, so we mentally build in an interval of uncertainty.

Now imagine that you have to make the same trip, but your grandmother has told you she'll write you out of her will if you don't arrive within ten hours exactly. How do you feel? How do you drive? Do you take time to stretch, or get a cup of coffee, or go a little slower through an icy patch? Of course not! You drive like a bat out of hell—as quickly as you can—because your financial well-being depends on it. Under this much stricter regime, even one minute past the deadline means you're in violation.

Truckers, of course, feel these same pressures based on the inflexibility introduced by the electronic monitor, with consequences for the safety of the public roads.[61] In one OOIDA survey, 78 percent of truckers reported feeling *more* pressure to drive when they felt it was better to stop—and 69 percent felt more pressure to drive in unsafe road conditions—than before the ELD mandate.[62] And under the mandate, while citations for hours-of-service violations dropped, citations for speeding and other forms of unsafe driving increased dramatically across all firm sizes; most notably, independent owner-operators saw a *35 percent* increase in unsafe driving violations after they had to install ELDs.[63] The researchers conducting the analysis chalked the spike up to "drivers attempting to make up for productivity lost" because of the stricter monitoring.[64] As one OOIDA rep put it: "[Truckers] feel controlled by that circle-thing on the dash, and they know they can't be a minute late. I've talked to thirty-year drivers, and they say they know that they shouldn't be driving that fast, but they are so worried they're going to run out of time."[65]

———

From the perspective of the federal government, the ELD is a regulatory tool, deployed for the purposes of improving highway safety (in spite of limited evidence that it does so). But the ELD is not *only* a regulatory tool—it is also a managerial one. In the next chapter, I consider how its use transforms relations between truckers and their employers, and how firms use the ELD to further their own business goals.

4

The Business of Trucker Surveillance

As data-gathering sensors and analytics platforms continue to become lighter, cheaper, and more powerful, more workplaces are using these technologies to manage and incentivize employees' performance at a fine scale. Trucking is no exception. Remote monitoring systems change how much firms know about workers' activities, and when they know it, as day-to-day work behaviors are made more visible and measurable.

But *how*, exactly, does this data collection get transformed into control over workers? How do electronic monitoring systems reorient information flows, and how are those flows used in managing workers? Data collection does not automatically or immediately change how people do their work or how managers oversee them. Firms have to develop specific managerial strategies to turn information into control. And in trucking, the pathways managerial control takes can be surprising, as firms put truckers' data to multiple and seemingly incongruous uses.

The trucking workplace was long immune from close managerial oversight, owing to its distributed and mobile nature: for most of the profession's history, it was simply impossible for firms to keep close tabs over their workers or to give them detailed instructions about how to do their work. Because of this, truckers retained more day-to-day occupational autonomy than many of their blue-collar brethren in more traditional work settings.

But this has quickly, and drastically, changed—aided in no small part by the federal ELD mandate.

Digital monitoring leads to two important dynamics that change how truckers are managed by their companies. First, monitoring *abstracts* organizational knowledge from local and biophysical contexts—what is happening on the road, around the trucker, and in the trucker's body—to aggregated databases, and provides managers with a trove of evidence with which to evaluate truckers' work in new ways and challenge truckers' accounts in real time. A trucker has traditionally had near-complete autonomy to move a load from point A to point B in the way he saw fit—an independence grounded in the fact that the trucker had exclusive knowledge of road conditions as well as the biophysical self-knowledge to know when he was too tired to drive. But since monitoring technologies transmit data about these conditions (or approximations thereof) to firms in real time, managers now use them to issue more fine-grained directives about what a trucker should do, as well as to construct a counternarrative of conditions that may contradict the driver's—diminishing both his autonomy and the value of his "road knowledge."

Second, firms *resocialize* abstracted monitoring data by reinserting them back into social contexts and creating social pressures on truckers to comply with organizational mandates. Firms gamify and encourage comparison between truckers in a fleet in order to foster competition for the best ratings, sometimes backed by a modest financial reward or public recognition. These processes capitalize on truckers' relations with their coworkers—or even their families—in order to pressure them into compliance with the goals of the company.

Even beyond these management dynamics, data collected about truckers can be a source of value for other business interests. Electronic monitoring has prompted "a gold rush" as third-party companies find ways to turn truckers' data into new forms of knowledge. Some of these data uses might benefit truckers, while others might impose additional costs on them. As we'll see, there are many ways in which surveillance is good for business.

The Data-Driven Workplace

As we discussed in chapter 1, worker surveillance is nothing new. Information technologies have long been used by managers to afford them control over their workers.[1] They do so by giving managers greater visibility into work tasks, allowing them to more precisely measure, direct, and incentivize productivity in accordance with the firm's aims—and to discipline workers if they don't comply.[2] These motivations and strategies date back at least to the rise of industrial production—but in recent years, new types of tools to help managers watch workers have proliferated. These tools may help firms capture more finely grained information (for example, logging every keystroke an office worker makes); they lower the costs of observation, as they become more commonly integrated into workplace software systems; they facilitate knowledge about new dimensions of worker behavior, like biometric data on workers' exertion or fatigue; and they enable monitoring of new workplaces, including those that have traditionally been difficult or impossible for managers to supervise. As we'll see, each one of these dynamics is at work in trucking.

Much work—including trucking work—has traditionally depended centrally on the use of the worker's body, the perception of cues in their immediate environment, and the accumulation of embodied knowledge as the worker learns by doing.[3] Even contemporary professionalized work often depends on a human being's direct observations and perceptions, as Phaedra Daipha documents in the context of weather forecasters[4] and Diane Bailey and colleagues show in the case of auto engineers.[5] Information technologies can attenuate this connection between work tasks and embodied knowledge—by breaking up work processes into discrete, rationalized, lower-skill tasks; by decontextualizing knowledge from the physical site of labor to centralized, abstract databases; or by converting work practices into ostensibly objective, calculable, neutral records of human action.[6] In these ways, information technologies legitimate some forms of knowledge while rendering others less valuable, to potentially detrimental effect on worker power.

A Legal Tool with a Business Rationale

Before ELDs were legally mandatory for all truckers, they were already being used voluntarily in some parts of the industry: though definitive numbers are tough to come by, some experts estimated that 30 to 50 percent of truckers were already using ELDs in the few years just prior to the mandate. This adoption was heavily skewed toward larger trucking firms. Why? It may have been that these companies were simply reading the writing on the wall, getting systems in place before they were required to. But a much more likely reason is that ELDs are not only useful from a regulatory perspective—they are also incredibly important business tools for large companies.

It may not seem obvious why large trucking companies would find value in adopting a new way to keep track of drivers' time—and one that in many ways makes the work of trucking less flexible for all drivers. But the ELD mandate was, for large firms, good for business. This is because—just as a truck driver does many things besides drive a truck—in many cases, an electronic logging device doesn't *only* do electronic logging. Rather, time logging capabilities are very often *bundled* with other performance monitoring and analytics functions—the same hardware unit captures all kinds of information other than timekeeping, and the same back-office software gives managers access to those data. The ELD is often integrated into what trucking firms call *fleet management systems*. In practice, then, the time-tracking functions of ELDs are very commonly hours-of-service *applications* bundled within fleet management systems, rather than discrete devices.

The range of information captured is quite extensive. It commonly includes a driver's fuel efficiency and idling time, speed, geolocation and geofencing (notifying a dispatcher if a truck has departed from a predetermined route or arrived at a terminal), lane departures and braking/acceleration patterns, freight status (such as the temperature of a refrigerated trailer), tire inflation, and vehicle maintenance and diagnostic information. In addition to monitoring and transmitting this performance data, fleet management systems typically contain additional modules that provide services like routing and two-way messaging; some units also integrate with toll bypass systems.[7] Other systems integrate with dashboard cameras—some that face outward to capture footage of the road, and others that face inward, recording the driver and the cab.[8]

Data from these systems can also provide a basis for predictive analytics—for firms to make predictions about which drivers are crash risks, or who is likely to quit. The proprietary Omnitracs Driver Retention Model analyzes hours-of-service data in order to identify "drivers suspected of flight risk"—which it does by looking for "patterns and subtle changes in driver habits and work activities that serve as indicators of voluntary terminations [quitting], such as the number [of] hours working, delays at customer sites, lack of hours, and amount of activity on the clock."[9] Omnitracs proposes that managers can use these data to initiate conversations with drivers and address problems. As one industry technology consultant explained to me, small changes in work patterns might indicate that the driver is struggling; the example he used was that if a driver who always started his day early suddenly began starting later, it could be a sign that he was going through some marital troubles.

It is these sorts of analytics and performance monitoring features that are most attractive and useful to large trucking firms, and that are the focus of much of this chapter. Trucking firms use these systems to manage truckers in new ways, as we'll see—and to provide "supply chain visibility" to customers who want to know exactly where their deliveries are and when to expect them. But the legal mandate remains an essential part of the story. The mandate provides a business rationale for firms to monitor driver behaviors not covered by the law: since they will be required to install ELD hardware in their trucks anyway, the marginal expense of also monitoring a swath of other driver behaviors using the same technological system is greatly reduced. Even FMCSA has recognized that "companies use [ELD devices] to enforce other company policies and monitor drivers' behavior"[10] for purposes *other than* compliance with the timekeeping rules.

As an analogy, we might imagine that FMCSA had created a rule requiring that everyone purchase a cell phone that makes calls. But of course, it's almost impossible to buy a phone that *only* makes calls, since most phones bundle call-making capability alongside many other standard features—so even though the regulator didn't explicitly *require* everyone to buy a cell phone that texts and accesses the Internet, the rule would end up having such an effect. The same applied to many truckers being required to install ELDs. (As we'll discuss in chapter 6, a small crop of "dumb" ELDs—systems that quite deliberately lacked these auxiliary monitoring capabilities—also arose in response to the mandate, and are used by some truckers as a form of resistance.) Representatives from the owner-operators' association advocated for the mandate to *require* "dumbing down" ELD devices by unbundling functionalities—that is, to require that government-mandated ELDs *not* contain auxiliary fleet management features. But this suggestion was strongly opposed by others, on the basis that it would be economically inefficient for firms to have to purchase separate devices.

Feature bundling is not the only way that the ELD became a business tool as much as a legal one; the precision of monitoring also comes into play. The ELD regulations don't require a device to monitor drivers' real-time exact location, but instead require only that the device record a truck's location at least once every sixty minutes. But in practice, the great majority of ELD devices on the market record and transmit location constantly via GPS tracking. As such, the realized surveillance capacity of the ELD far exceeds what's technically required by the legal rule.

There are also competitive forces at play. Large trucking carriers were initially skeptical about mandatory ELDs, and said so in public comments to proposed FMCSA rules. But as the monitoring technology improved, their opinions changed. In the final rulemaking process, the American Trucking Associations (the largest trade association in the industry, representing the biggest trucking companies) spoke out strongly in favor of the devices, touting their potential to reduce fraud and improve safety—while small carriers and owner-operators continued to strenuously oppose them. This change of heart was no doubt motivated in large part by the value large firms were coming to see in the ELD mandate. Many in the industry believe that smaller carriers also flout the timekeeping rules more regularly than large firms—so an across-the-board crackdown might "level the playing field" to larger firms' benefit.

ELDs and the data they collect offer much more value to large firms than to their smaller competitors, allowing large firms to more easily recoup their investments in the systems (while imposing costs on smaller firms, who will not reap those advantages). One safety study explained this "structurally asymmetric impact" on firms of different sizes in this way: "The monitoring and informational advantages of ELDs are reduced for small carriers, whose operations are less complex compared to large carriers, and the benefits are practically non-existent for independent owner-operators (except for a reduction in paperwork)[.]"[11] Smaller players tend to "self-monitor" and lack the principal-agent problems that large companies can address through monitoring and analytics.

It should be noted that trucking firms face competing economic incentives regarding electronic hours-of-service monitoring. Companies don't want their fatigued employees to have accidents—but they *do* want them to transport goods quickly, and some may be willing to break the law to do so. For large firms especially, the potential loss of employee productivity can be economically offset by safety benefits, reduced litigation risk, and savings on insurance premiums and internal auditing costs—as well as enhancements to efficiency supported by performance monitoring. In some cases, firms try to have it both ways, instructing drivers about how to exploit the technical limitations of the monitor to evade the timekeeping regulations without being caught, as I'll describe in chapter 6.

How Electronic Monitoring Reconfigures Information Flows

Fleet management systems reorient the information that flows between truckers and their employers across three closely related, yet analytically separable, dimensions. They change the nature of what information matters, allowing firms to find value in data *aggregation* rather than truckers' bio-physical and road knowledge; they change the nature of who has information, taking away the *exclusivity* of information truckers previously had; and they alter the *temporality* of knowledge—who has information when. Taken together, as we'll see, these three shifts facilitate new management strategies by firms.

AGGREGATION

When fleet management systems are used to monitor truck drivers' work, *aggregated* information comes to supplant "road knowledge"—individual, local, and biophysical self-knowledge gleaned from years of experience—in terms of value to the firm. Traditionally, truckers' value, and a good deal of their occupational identity, arises directly from their deep knowledge of the daily "ins and outs" of driving a truck—from how trucking equipment works and how to fix it quickly when needed, to local conditions that affect their work (what roads are closed, what time of day traffic is the most onerous in various locations), to routing information (how long it takes to get from Omaha to San Francisco, and the best way to get there depending on weather conditions).

Truckers—especially "old hands"[12] with millions of driving miles under their belts, who have been on the road for years or even decades—take pride in this knowledge and show considerable pleasure in regaling one another (as well as the occasional researcher) with detailed stories of how it has benefited them in some manner or another on the road. Truckers often emphasize how this knowledge helps them to do their jobs without needing to use technological "crutches," like GPS, as support. Consider the following exchange I had with one driver about how to get from Oregon to Indiana:

Q: So you don't use GPS though?
A: GPS? No. Honey, I've been driving for twenty-nine years, I've been all over the United States, I don't need a GPS. I don't even need a map.
Q: You don't use a map?
A: [laughing] No.
Q: Really?
A: Hell, no. I could drive—where do you want to go?
Q: West Lafayette, Indiana. [...]

A: Go around Ontario, Oregon, over to Pocatello. Go south on Pocatello, go to McCammon, that's 30, it runs—McCammon runs over to 80, I-80, that'll come out by Little America, take Little America—or the 80, excuse me—run that over to Chicago, right? Get through Chicago, now from there it's up to you which way you want to go. [. . .] You'd have to go south on 65, down towards Indianapolis. [. . .]

Q: So how do you learn all this [about different routes]?

A: Honey, driving them.

Exchanges like this were not uncommon in my conversations with truckers, as were "war stories" about drivers' experiences navigating particularly knotty routes or demanding timetables over the years. Truckers also described ways in which their own knowledge trumped the ineptitude of other human beings, like dispatchers or inspectors, or the shortcomings of technologies themselves, like faulty routing equipment. Road knowledge serves as a crucial source of both economic and cultural value for these workers.

Pride in accumulated road knowledge extends, too, to bodily self-knowledge about how fatigued a truck driver feels. Recall from chapter 3 the entanglement between truckers' identities and their biophysical stamina; drivers are strongly resistant to being told what their own individual biophysical limits are. As one driver put it:

I'm not going to work under conditions where I'm treated like a child, a child who doesn't have enough [sense] to know when to go to bed and when to get up; or when to stop and rest while rush hour traffic clears and then proceed when rested and safer.

For many drivers, then, professionalism and occupational pride are deeply entwined with knowledge of biophysical and local conditions, which have long been of primary value in the effective completion of their work. But the value of these forms of knowledge has been displaced by fleet management systems, which can directly measure (or at least make actionable guesses about) conditions in and around the trucker. A trucker's knowledge of the best way to get from one place to another is of less use in the face of automatic routing and geolocation. Remote vehicle diagnostics obviate some of the need for truckers to know about truck mechanics. And of course, hours-of-service monitoring and fatigue management technologies dislodge the need (or even the capability) for truckers to drive when their bodies feel up to it and to stop when they need to rest.

Not only do these capabilities replace road knowledge and self-knowledge, but they introduce a new form of valuable information—aggregation and comparison of information about multiple drivers, gleaned from afar. A driver may state that a road is currently impassable due to weather conditions, an assertion that would previously have been difficult to challenge—but using aggregated data, his dispatcher may respond that "I know the weather is not *too* bad for you to continue driving down I-80, because I see that I have four other trucks on that road now." (This example and others in this chapter are drawn from my observations with dispatchers in trucking firms.) And as we shall see in a moment, back-office modules readily summarize drivers' information in order to create detailed performance metrics and comparative "scorecards" for drivers, based on whatever criteria they see fit; fuel economy, on-time delivery ratings, and safe driving practices are commonly used.

EXCLUSIVITY

A second dimension of change concerns *who* possesses valuable information. When monitoring systems are used, truckers are no longer the sole holders of relevant knowledge, as fleet management systems distribute it to remote dispatchers. Truckers themselves still possess knowledge of their work activities, but they are no longer its *exclusive* possessors, which has significant implications for their power to direct their own work.

Traditionally in trucking, a driver (even an employee driver) has been considered the "captain of his ship": like captains of other transportation vessels, the trucker holds the ultimate authority to stop driving if, *by his own judgment*, he is too fatigued to do so, or if local weather or road conditions make continued travel unsafe. Drivers were accorded a high degree of autonomy in deciding when and how to conduct their work, which is closely bound up with the emphasis on local and biophysical knowledge. As we discussed in chapter 3, for many drivers, this decision-making independence is what initially attracted them to the profession, and it serves as a strong source of professional pride. Prior to the proliferation of cell phones, drivers would share information with dispatchers from truck stop pay phones—when they chose to. Even after cell phones became common, truckers still had exclusive command over a good deal of important information about their location, conditions, and plans for executing their work.

Yet when fleet management systems transmit information from trucks to back-office dispatchers, firms have much more information at their disposal—information that conveys to the dispatcher knowledge about the driver's internal and local conditions, as well as aggregated data about drivers' performance. Under the "ship captain" model, a driver may declare authoritatively that he is too fatigued to drive; but when the driver's hours of service are monitored by a fleet management system, a dispatcher may respond to the effect that "I know you aren't (or shouldn't be) too tired, because I can see that you've only been on duty for five hours."

The importance of this shift is perfectly illustrated in the following exchange, which received some attention in industry media.[13] In this case, a driver on a legally mandated sleep break received the following string of messages from his employer, via his fleet management system's two-way messaging capability (emphases mine):

12:57 pm   Firm: Are you headed to delivery?

1:02 pm   Firm: Please call.

2:33 pm   Firm: What is your ETA to delivery?

2:34 pm   Firm: Need you to start rolling.

2:35 pm   Firm: Why have you not called me back?

3:25 pm   *Driver: I can't talk and sleep at the same time.*

3:37 pm   Firm: Why aren't you rolling? **You have hours** and are going to service fail this load.

3:44 pm   Firm: **You have hours** now and the ability to roll—that is a failure when you are sitting and refusing to roll to the customer.

3:51 pm   Firm: Please go in and deliver. We need to service our customers. Please start rolling. They will receive you up to 11:30. Please do not be late.

4:14 pm   *Driver: Bad storm. Can't roll now.*

4:34 pm   Firm: Weather Channel is showing small rain shower **in your area**, 1–2 inches of rain and 10 mph winds ???

A number of important things are occurring in this exchange. The first, of course, is the fact that it occurred at all. Not long ago, the firm would not have been able to contact a trucker on the road in the first place, but would have been reliant on the trucker to check in. Note, too, the frequency with which the dispatcher contacts the (ostensibly sleeping) driver to tell him to get back on the road: the messages come at one-minute intervals for part of the exchange. But even more importantly, the dispatcher repeatedly invokes his *own* approximated knowledge of the driver's biophysical status and conditions in his location ("you have hours"; "in your area"), derived from the fleet management system's back-office capabilities (his ability to see the driver's hours of service, and his use of the "Weather Channel"—by which he likely means a weather map overlay that dispatchers can use to check conditions around their trucks).

TEMPORALITY

Finally, fleet management systems facilitate a temporal shift in information flows, as information is distributed in real time. When truckers tracked their time manually, the regulations required him to keep his paper logs up to date at all times by updating the log each time his duty status changed (such as when he began driving for the day, stopped for lunch, or arrived at a shipper's terminal). But in practice, drivers were often well-behind on recordkeeping during a haul. There was often a significant lag between these activities happening and a driver recording them—and an even bigger delay between the driver recording information on his paper log and a dispatcher having access to that information. (Prior regulations required a trucker to submit paper logs to his employer within thirteen days of completion, though drivers commonly would submit logs upon returning from a trip.) This lag time gave the driver ample opportunity to make adjustments to the paper logbook before submitting it to his employer, which therefore gave him "wiggle room" to appear compliant in *post facto* records—whether or not his actual driving behavior was in fact actually in line with legal rules.

One driver described to me how the paper-based system afforded him flexibility:

Q: So, did you have to keep track of your time before you got the Qualcomm [ELD]?
A: Yeah.
Q: How'd you do it?
A: With a logbook. But in a logbook . . . it works pretty well, because you can chisel parts off here, and parts off there, and you make the whole trip fit perfectly.
Q: So, when you say you're chiseling, does that mean that you're twisting it a little?
A: Well, you do a fifteen-minute pre-trip [inspection]; then you get caught in traffic for, like, an hour. You can give up a meal break, you can give up an equipment check, and that covers that hour.
Q: So, you just fudge it a little bit.
A: That's it.

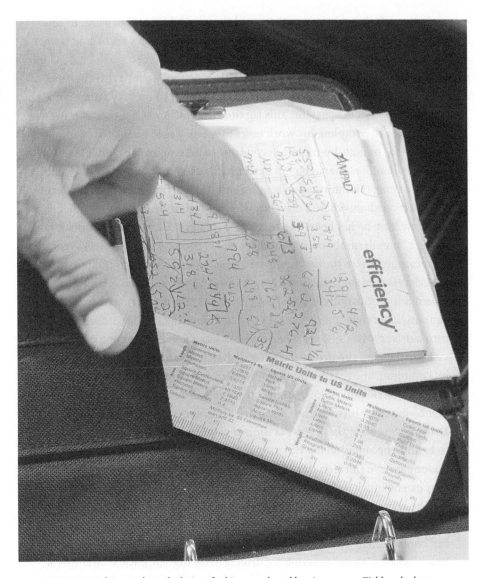

FIGURE 4.1. A driver makes calculations for his paper-based logging system. Fieldwork photo by Alex Zafiroglu.

When this trucker refers to "giv[ing] up" a meal break and equipment check, he means that he would *record* a rest break and equipment inspection in his paper logbook, but not actually take the time to perform these activities. Rest breaks and equipment inspections are legally required, and it is thus in his interest to report having completed them; the time it takes a driver to complete them counts against his fourteen-hour daily work limit but *not* the crucial eleven-hour daily driving limit. By "chisel[ing]" these activities off, this driver uses the saved time for additional driving. In this way, he would gain an extra hour of driving time without recording it on his log.

Under the paper logbook system, then, it is common for a driver's time-keeping records to be reconstructed after the fact to portray compliance regardless of actual behavior. This lag time thus becomes a source of power for a driver to complete his work tasks more or less as he sees fit, without the need for minute-by-minute accounting to his employer or other parties.

But when fleet management systems are used to monitor a driver and track his time, this information is transmitted in real time back to employers. Information about where a driver is, whether he is moving and how quickly, and how long he has been on the road or on a break is constantly updated. Using a "breadcrumb" map view, dispatchers can watch all of a fleet's truck drivers move across the country in real time. All drivers' to-the-minute drive time data is viewable in an orderly spreadsheet as it happens. Not only is this information transmitted in real time, but it is also predictive: dispatchers can anticipate violations *before* they occur. Some systems' dispatcher portals turn a driver's data cell red when he is within a few minutes of a violation, visually flagging the driver so that a dispatcher can respond immediately by communicating with and directing him as to his next course of action. This may be helpful—like helping a trucker find a safe place to park for the night—or it may involve pressuring drivers to speed up to avoid a violation. Drivers accordingly have less decision-making autonomy to determine how to handle the situation if a violation is about to occur. And because drivers no longer have thirteen days to strategically "correct" a log after the fact, they can be much more readily penalized by their employers for violations.

The temporal shift can also enable harassment of drivers by employers. Even when drivers are off duty, employers can see where they are, and can contact them using systems' communication functions—which sometimes lacked a "mute" function for drivers to silence employers' attempts at communication, even during sleep breaks. (FMCSA regulations eventually required that ELDs have a mute function to ensure that drivers could sleep uninterrupted.) One driver told me that other drivers in his fleet took steps like removing fuses to prevent being contacted during off-duty hours, but that this was a risky, fireable offense (see chapter 6 for a detailed discussion of such practices). If, for example, a driver has not left a rest stop as soon as it becomes legal for him to do so, a dispatcher may call and pressure the driver as to why he is not on the road yet (as in the "Weather Channel" exchange discussed in the previous section, in which the dispatcher sent repeated pleas to an ostensibly sleeping driver to "roll" at one-minute intervals). In one industry survey, 68 percent of drivers reported being told by firms to drive longer, and 29 percent reported being awakened to be given these instructions.[14] Truckers are legally protected against harassment via ELDs—though the bar for what constitutes harassment is quite high; it only encompasses actions taken by firms that induce workers to violate the hours-of-service rules. (The "Weather Channel" exchange, for example, would not likely constitute harassment under the FMCSA's definition.)

New Information Flows, New Management Strategies

ELDs and fleet management systems therefore bring about important changes in what information is valuable, who has it, and when they have it. But how do firms put these new information flows to use, and how do these strategies give them new forms of control over truckers? Firms use abstracted data to evaluate truckers in new ways and challenge truckers' accounts about local and biophysical conditions. They often subsequently resocialize abstracted data by strategically deploying it into truckers' social relationships, including their family lives. I'll discuss each in turn.

ABSTRACTION: EVALUATING PRODUCTIVITY
AND CHALLENGING ACCOUNTS

Fleet management systems restructure organizational information flows by reconstituting truckers' embodied work as a set of divisible, rationalized data points, presented in an apparently neutral format. These data are divorced from the context of road conditions, the contingencies of weather and shippers' schedules, and other individuated circumstances.

This abstraction has two important effects on drivers' occupational power. First, it facilitates quantitative performance evaluation when such data are recombined as driver scorecards. On the basis of these metrics, managers can easily make comparisons across different periods of time, both within and across drivers, across groups of drivers (as defined by back-office managers on any basis: managers can compare customizable driver groups based on type of equipment, type of haul, driver experience level, or any other imaginable axis of variation), or between a company's fleets and industry averages. This aggregated information, summarized in quantifiable and easily comparable metrics, becomes a highly valued management tool.

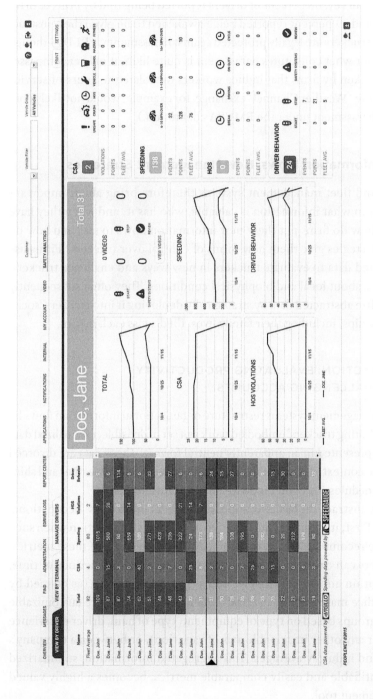

FIGURE 4.2. A fleet safety analytics dashboard and driver comparison chart. Credit: *Overdrive*, https://www.overdriveonline.com/business/article/14890468/peoplenets-video-system-analytics-dashboard.

Second, real-time digital monitoring provides companies with evidentiary fodder for challenging truckers' own accounts of local and biophysical conditions. This reduces firms' dependence on truckers' self-reports and may give them more control over truckers' labor. As described, a dispatcher can consider a trucker's declaration that a road is impassable *in the context of what he now knows* about how other drivers are performing under the same conditions, or can consult a driver's historic hours-of-service records or fuel-usage scores in appraising the reliability of his statements about extenuating circumstances that lead to a violation of safety regulations or company policy.

The "Weather Channel" exchange makes this use of data abundantly clear: recall that in that exchange, the driver's assertions—that he needs to sleep, and that the weather is too poor for him to drive—are directly countered by the dispatcher's *own* assessments of the situation via back-office software. The dispatcher has broad but indirect knowledge here: they have fine-grained, real-time digital data about every truck and trucker in the fleet, though these data may be poorer approximations of what they purport to measure (fatigue, safety risk, etc.) than what the driver can perceive. This is a common tension in the use of technology in geographically distributed work: managers who are not "on the ground" where the work is occurring often have imperfect, attenuated views of the situation, and may or may not draw accurate inferences as a result.[15]

Importantly, this is not to suggest that truckers' accounts are always entirely accurate or forthright, either. In the "Weather Channel" exchange, for example, the driver may or may not be telling the truth about local conditions or in implying that he needs to sleep; the tone of the exchange is clearly confrontational. But the veracity of the trucker's account is not the point: the point is that digital data facilitates this challenge between the trucker's account and the dispatcher's, in which each is marshalling different forms of information as evidence to support their own version of events.

Many drivers read this use of data as evincing a lack of trust. As one driver put it: "If you can't trust me to go out there and be safe and honest, then take me out of the game and put somebody in there that you think can. Either that or put a robot in the truck!" Others feel that managerial reliance on abstracted data is utterly incompatible with their highly contingent, unpredictable work:

Q: So, what is it that you don't like about the [ELD]?
A: There are things that happen in the life of trucking that can't be overcome.

Q: Like what?

A: Accidents on the road, breakdowns, flat tires—anything that will slow you up. From Point A to Point B it's, say, two hundred miles. You've got four hours to get there. That's fifty miles an hour, that's a good average for a truck. If anything happens between that—a wreck, anything where you're just sitting in the middle of the road wasting your goddamn time—excuse me, your time—you can't make that. And then they want to know why not.

Q: Who's "they?" The company?

A: Yeah. And the customer especially. Well, I've been driving—like I said, I've been driving truck forty-two years. I'm not in the habit of explaining myself. Having me to explain myself is really an insult.

Q: So, how does the computer affect that?

A: Well, if I've got to sit there for forty-five minutes or an hour, I'm down to three hours delivery time. It don't give a damn if I'm sitting there in a wreck.

Q: The computer, you mean?

A: No, it don't know nothing about that.

This sense extends from local to biophysical conditions as well, as another driver put it: "[Y]ou, as a professional, you know when your body is tired. You know when your mind is fatigued. You know when you need to stop and rest. That dispatcher doesn't know. And by God, that electronic device certainly does not know."

Of course, the fact that digital data make it possible for dispatchers to challenge drivers' judgment does not mean that they *always* do so. The effects a technology has on workers are largely contingent on the organizational practices adopted by companies in implementing it.[16] Based on my observations in trucking firms and on what drivers shared with me about their experiences, there is significant variation in the degree to which firms question drivers' judgment based on the data at their disposal, particularly when safety is at stake; many drivers still feel like they have ultimate authority to make the decision not to drive if they feel doing so would be unduly risky. However, it is also likely that the fact that data are being collected—the fact of which drivers are very much aware—prevents drivers from making claims that the data would not support: it is difficult, for example, for a driver to say he is out of driving hours in order to avoid an undesirable load assignment, because he knows that those data are immediately accessible to dispatchers and trip planners.

Interestingly, one trucking firm I spoke with purposefully de-links hours-of-service data from the software used by their dispatchers, so that "drivers have to tell us how they are managing their hours and the next time they will be available for dispatch. . . . Bottom line is our drivers are the captain of their ship. They tell us what they can do, not the [ELD] data." But this situation was unique in my discussions with firms and truckers; in most cases, it appears that dispatchers are given ready access to information about drivers' locations and legally available hours in order for these data streams to be used as managerial tools.

RESOCIALIZATION: SHARING DATA TO CREATE SOCIAL PRESSURE TOWARD COMPLIANCE

Another important pathway toward organizational control of truckers is firms' deployment of digital data into truckers' social lives, both within and outside of the fleet, in an effort to foster social pressure toward conformity with organizational goals.

It is not at all unprecedented for truckers to share information within their professional community. Drivers frequently share information among themselves on online forums, in informal conversations at truck stops, and in other venues—they share war stories, best practices, tips for avoiding law enforcement, and the like. These exchanges build community and social solidarity among drivers, support their occupational identities, and encourage the formation of professional pride.

But unlike these knowledge-sharing activities, fleet management systems enable information about truckers' activities to be shared, without their direct agency or consent, in order to create competition among workers. Many fleet managers post or distribute rankings based on drivers' scorecards, which the systems make technically very easy to produce. Scorecards display driver safety records, hours of service, or other performance indicators that align with organizational goals; fuel efficiency is a very popular metric, given the high present price of fuel and its significant impact on trucking companies' financial bottom lines. The ability to easily aggregate, specify, parse, and compare this data across multiple drivers is one of the chief advantages of such systems, particularly for larger companies.

By posting performance data where drivers can see it, companies create social pressure for comparatively underperforming drivers to improve and compete.[17] A system trainer (who advises firms about how to adopt and use

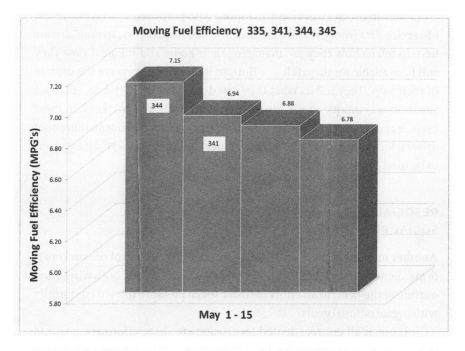

FIGURE 4.3. A driver analytics comparison chart posted in a trucking company common area at which I conducted fieldwork. This chart uses drivers' identification numbers (rather than names) and compares their fuel efficiency.

fleet management systems) described to me how such displays are utilized to motivate the least efficient drivers in a fleet:

> [Companies] just put the list up and it would say, Driver A, his MPGs are 8.3, and Driver B is 8.2, and Driver C is 7.6, and Driver D is 6.2. And you post it up. Well, if you do that, what is it going to tell you? I don't want to be the one at the bottom, and next week the butt of the jokes. So I [the company] might not even say it, I just show it. And now that guy tries to do a little bit better. [. . .] So it all depends on how you want to incentivize your drivers [. . .] some of it is just posting it up on the wall and letting them look at it.

This strategy, then, depends on drivers feeling interpersonally shamed by their coworkers ("the butt of jokes") as a result of inefficient performance. Vendors describe these kinds of data-driven gamification as ways to "engage and inspire drivers to do their best."[18]

Many firms go further by directly tying small financial incentives to the rankings produced by fleet management systems. These incentives often cost firms very little, but coupled with the pride (or embarrassment) of one's comparative ranking, were considered effective tools for aligning the goals of firms and employees:

[You're] a fuel company, let's say, and you want your drivers to improve. And if they get a score of let's say 90 or above, you give them a 25-dollar gas card from your company. Well, not only are you giving them 25 bucks, they think it's great, but it's not costing you 25 because they're buying your gas! [. . .] We had customers that did that all the time. And it wasn't just gas, but other things, because we had customers that did food [. . .] and the drivers could go eat at the restaurants that they service . . . so it's not really costing you 25 bucks. It is, but it isn't. And drivers appreciate it and they start to do things a little bit better. You know, and it doesn't have to be a lot anyway. So that's why a lot of people do it that way.

These incentive alignments, though often substantively small for companies, can provide powerful symbolic motivations for efficient employee performance. Incentivizing employees through intra-workforce competition is not new;[19] yet, the degree to which these programs comprise a fundamental element of fleet management is striking. Technical trainers (who work for the technology vendor) counsel trucking firms on how to integrate such schemes into their use of the technology, often under the rationale of "culture change," just as they provide advice on the technical workings of the system. Marketing materials and vendor-issued white papers stress the importance of rolling out incentive programs along with the installation of the systems, advising managers that "[b]arriers to adopting telematics solutions are usually about how people accept change. . . . It is possible to drive improved business results using people's natural inclination to be competitive."

Firms' efforts to resocialize monitoring data do not end within the company. Firms also invoke social pressures in drivers' own *non*-trucking communities as well, particularly among their families. Incentives like awards ceremonies and banquets, to which drivers' families are invited, are common strategies. But involvement can be even more directed, as well: for instance, one firm sends small bonus checks for the highest-performing drivers (as determined by driver scorecard data) to the drivers' *wives*, in the wives' names. The idea behind the program is that wives come to expect the checks periodically (as "a profit-sharing arrangement," as it was described to me, ostensibly in recognition of a wife's familial support of her trucker husband); wives are expected to create pressure for their husbands to continue meeting the company's organizational performance benchmarks. Recalling the strong "family man" masculinity many truckers exhibit, it is unsurprising that firms' control techniques capitalize on this normative orientation toward economic provision for one's family. But it is notable that here, data-driven social control still depends very much on being situated in "soft" social forces like familial care—contrary to the common narrative of data-driven management being abstracted and impersonal.

The Values of Data

As we can see, the ELD is not purely a technology of compliance; it's also a tool for organizational control, as trucking companies use ELD data to try to align their workers' incentives with their own profit motives. But this isn't the only way companies find value in ELDs. In the aggregate, the data these tools capture is incredibly valuable, prompting what industry media have called "a gold rush for ELD data."[20] Some ELD vendors' terms of service allow for data sharing and uses that go well beyond the organizational and regulatory functions we've discussed so far.[21] Though a few vendors make it possible for companies to opt out of selling or sharing data outside the company, many don't—and in any case, any decision to opt out is likely to rest with the company, not an individual trucker employed by that company.[22]

Insurers, for example, find ELD data quite valuable, and sometimes offer users substantial discounts if they share the data. Trucking insurers have faced losses in recent years in the face of more distracted driving, more crowded roads, and less experienced drivers behind the wheels of big rigs—ratcheting up premiums and prompting insurers to find new ways of mitigating risk.[23] New usage-based insurance programs for truckers, like Progressive's "Smart Haul" program, offer a discount on premiums based on their driving data, with lower prices for safer driving. These programs are increasingly attractive to insurers who see them as routes to more accurately predicting risk. And more generally, even if they don't look at the data themselves, commercial insurers may offer better rates to companies that *install* camera systems and other data collection infrastructures—under the assumption that doing so will translate into lower risk of accidents (or at least, reduced liability).[24]

Even for companies not making direct use of truckers' ELD data, the ELD requirement can open up new lines of revenue. Truck parking was a scarce resource even before the mandate—and after the mandate, as drivers lose flexibility as to when to stop, it can be even harder to find lots with space. (The safety of parking is also an important issue; truckers who park in unsecured lots can be more vulnerable to robbery and even violent crime.) Truck stops have traditionally offered up spots on a first-come, first-serve basis, but the ELD mandate has pushed truck stops to offer more reserved parking, which can be booked online for a fee—roughly fifteen dollars per night. (Only about 15 percent of trucking firms reimburse for these fees.[25]) It has also helped spur the use of third-party apps like Trucker Path and private truck parking facilities.[26] On one hand, services like these can be helpful for truckers, who routinely spend time (and therefore money) hunting for scarce parking—but others argue that it just provides yet another way to squeeze truckers for money.[27]

There are some secondary uses of ELD data that may be advantageous to truckers. For example, some companies use location and hours-of-service data for load matching, a process through which some truckers look for nearby shippers with freight to move, and vice versa (rather like Uber's matches of drivers to passengers). Load matching is typically executed through brokers and third-party "load boards" that operate as middlemen to the match; ELD data are beginning to be integrated into these platforms, and could make the matching process more efficient.

In addition, some see promise in the use of aggregated ELD data to advocate for truckers' interests. The American Transportation Research Institute has established an opt-in ELD Data Clearinghouse for research purposes, and has used ELD data to advocate for changes to the hours-of-service regulations, demonstrating how the strictness of the rules (coupled with digital enforcement) forces truckers to drive during congested periods, rather than waiting until traffic clears.[28] Others have used ELD data to help predict maintenance needs and to gather "collective intelligence" on road conditions—for example, the company Geotab used ELD accelerometer and swerve pattern data to build a "pothole detector" it makes available to transportation authorities.[29]

ELD data might also have a role to play in helping truckers fight against detention time—unpaid time spent waiting to be loaded or unloaded at a shipper or receiver, discussed in chapter 3. Mandatory electronic monitoring for all truckers incidentally illuminates just how bad the detention problem is, and which shippers and receivers are the worst offenders. This is because though ELDs primarily surveil truckers, they incidentally offer a window into what shippers and receivers are doing, too (a phenomenon called *refractive surveillance*[30]): we can safely assume that if a trucker's ELD records him sitting idly or making small yard moves at a shipping terminal for hours on end, he is there because he is being detained by the customer. ELD data can thus provide "hard evidence" of a trucker's wait time.

KeepTruckin—an ELD company favored by owner-operators, who tend to have less power to push back against detention time or to put detention fees in their contracts—anonymizes and pools data from hundreds of thousands of trucks using their devices. It then uses this data to track delays at docks across the country and offers this data back to truckers through a free service called Facility Insights. The goal is to empower truckers to know *before* accepting a load whether they are likely to spend a long time waiting at the dock—and to pressure shippers to minimize wait times based on this visibility.[31] KeepTruckin also used these data to advocate for truckers in Washington; analyzing a representative sample of its users' data, it submitted a study to the DOT showing that its users were detained more than two hours about seven times each month—and that drivers drive faster (and, presumably, less safely) after a detention period because of their need to make up for lost time.[32] KeepTruckin argued to the DOT that detained truckers should receive two hours of extra work time on their clocks when they face extended detention, in order to avoid having to race the clock.

Interoperable Surveillance

The proliferation of electronic monitoring upends truckers' relationships with the companies that employ them. The systems make truckers' work—previously immune from immediate oversight due to their mobility and spatial distribution—newly visible to the firm by reconfiguring what information is valued, who has it, and when. These information flows provide fodder for the evaluation of truckers' performance with respect to one another, in terms of newly quantified metrics, and a basis for challenging drivers' accounts. The strategy is given extra "teeth" when data are strategically re-embedded in truckers' social networks of coworkers and families to compel compliance with the firm's objectives. The trucking case reveals new pathways through which workplace surveillance can entrench managers' power over workers. Even owner-operators—who work only for themselves—are implicated by this change, as the ELD becomes a competitive tool that augments the economic position of large carriers relative to smaller ones.

But the story neither begins nor ends there. It begins not with companies, but with the government. The FMCSA goes out of its way to note that the ELD mandate is concerned only with enforcing the hours-of-service rules, that it does *not* require the collection of performance data like truckers' fuel efficiency or speed patterns—and that the government doesn't access this data for its own enforcement purposes. But the government's requirements still provide a crucial backbone for corporate surveillance by scaffolding the collection of fine-grained performance data. And on the other side of things, trucker surveillance facilitates new uses of the data by third parties, creating value propositions for actors like truck stops and insurers.

We might think of government, corporate, and third-party surveillance as being fundamentally interoperable—while they are accomplishing different aims, they are deeply compatible with one another, and each facilitates and supports the others.[33] This interoperability is partly *technical* in nature: the ELD is a singular "black box" that bundles each of these dynamics together. The regimes are *economically* interoperable, in that ELDs provide a business rationale for firms to monitor driver behaviors beyond those required by the regulation, and serve the competitive interests of large firms.

Government and corporate surveillance are also *legally* interoperable: despite FMCSA's prescription of only limited data-collection requirements in the rules, FMCSA explicitly acknowledges companies' use of ELD data for performance monitoring and implicitly recognizes that many truckers will fulfill the legal requirement through purchase of a much more full-featured system. (For instance, in some of its proposed rules, FMCSA based its cost/benefit calculations on the adoption of the Qualcomm [now Omnitracs] MCP-50—a relatively low-end system, but one that is compatible with a wide range of auxiliary monitoring capabilities.[34]) The agency used the MCP-50 as its cost benchmark due in part to the fact that Qualcomm had the largest North American market share in fleet management systems at the time. Though the rule wouldn't have required the purchase of the MCP-50 in particular, its use as a standard suggests legal recognition of the reality that, in practice, legal and corporate surveillance interests are deeply intertwined.

And finally, government, corporate, and third-party data collection are *socioculturally* interoperable. The confluence of multiple forms of surveillance in one unit means that they can be quite difficult for drivers to disentangle. In my conversations with drivers, it was not uncommon for them to conflate the data collection directed at them by their employers, by insurers, and by the government—for example, believing that the government had automatic "back-office" access to all data collected through an ELD. This conflation is understandable, of course, given the union of all these data flows within a single artifact and the lack of transparency about data transfer and re-use.

Though I have separated these categories analytically, in practice they overlap inextricably and feed into one another. Firms' economic interests influence their inputs into regulatory processes; technical standards become instantiated in legal rules. Legal, sociocultural, economic, and technical interests are tightly coupled and mutually constitutive in the ELD case. Taken together, these synergies make public and private systems of surveillance pragmatically inseparable. The layering of surveillant interests through interoperable systems results in a greater net monitoring capacity than that which either the state or the corporation might achieve on its own, creating a hybrid, mutually enforcing assemblage of surveillant regimes.

5

Computers in the Coop

As we have seen, the emergence of digital monitoring in trucking fundamentally changes how truckers accomplish their daily work and what expertise they can draw on in doing so. But monitoring technology changes the game for other groups of workers, too. Here, we look at how the ELD changes the work of law enforcement officers tasked with implementing the timekeeping rules on the ground. Specially trained law enforcement officials—often members of state police departments and highway patrols—work as commercial vehicle inspectors, enforcing the myriad rules and regulations to which truckers and trucks are subject, either at roadside or in inspection stations that truckers are required to drive through when they are open. As electronic monitoring proliferates in the industry, these law enforcement officers must interact with and rely upon ELDs in the course of their daily work inspecting truckers' time logs. And as with truckers, the introduction of digital tools into law enforcement officers' work has not been seamless—it has disrupted their long-held routines and practiced modes of interaction. In this chapter, we explore what this transition has looked like and the effects it has had on truckers' relationships with law enforcement officers.

The story of the ELD's introduction into law enforcement work is critical for understanding how people work with, around, and sometimes against technology in their workplaces. Policing is, of course, itself a form of work, the dynamics of which can be disrupted by the introduction of new technologies. As scholars like Sarah Brayne, Bryce Newell, and Lauren Kilgour explain in their studies of policing technologies,[1] we cannot fully understand the impacts of policing technologies on the public unless we also understand how law enforcement officers conceptualize, use, and contest these tools—including how new tools challenge officers' own work practices and routines. As Brayne points out, this requires an inversion of our typical lens for thinking about state surveillance, in which we tend to view police as surveillors, rather than surveilled.[2]

In the trucking context, by understanding how law enforcement officers use technology and how their interactions change around it, we can uncover an underspecified source of bias in the outcomes that digital monitoring regimes seem to bring about—and a reason why we might question the quantitative measures that we use to evaluate these systems' effectiveness. One way we might measure whether ELDs are having their intended effect—that is, reducing the hours truckers drive to comply more closely with what the law requires—is by measuring the number of citations that law enforcement officers issue to truckers for violating federal time limits. Some evidence suggests that before the mandate, drivers who were digitally monitored were less likely to receive these citations than were drivers logging via paper and pencil. And in the first six months after ELDs were made mandatory for everyone, the number of hours-of-service violations reported by the FMCSA was cut roughly in half.[3] Thus, ELD proponents argue, this form of electronic surveillance prevents truckers from driving past the point of fatigue; digital monitoring achieves its intended goal.

But it's hard to make sense of these statistics. On one hand, it could be that ELDs *do* actually compel drivers to violate time limits less often. But looking closely at interactions *around* the monitors, between law enforcement officers and truckers, suggests an alternative story that might explain at least part of the reduction in citations issued to monitored drivers.

Law enforcement officers' inspection processes are very different when a trucker is using paper versus electronic logbooks. Officers have had decades to develop protocols for inspecting paper logs. And as we shall see, officers inspecting paper logs do so in such a way that puts them in a position of power and authority vis-à-vis the trucker, by making strategic use of time and distance during the inspection. In contrast, officers may have considerable difficulty acquiring data from electronic monitors—and these technical troubles are overtly visible to drivers, thanks to their location within the truck cab. When officers inspect electronic monitors, their perceived authority and expertise may be undermined, potentially leading them to conduct fewer and less rigorous inspections—and, therefore, to issue fewer citations, thus potentially biasing the differential outcomes that result.

As such, the monitor counterintuitively destabilizes traditional power dynamics in law enforcement officers' interactions with truckers by undermining officers' expertise and confidence. The case suggests that, to understand data-intensive systems, it is important to recognize not only how bias can be "baked in" to algorithmic processes (through modeling assumptions, training data, and the like)—but also how technological black boxes, *once created*, operate in broader social environments. Actors must negotiate around these new tools in order to preserve their own social and professional interests. Any effort to create accountability for such systems is incomplete if the inquiry stops at the perimeter of the black box. Inquiries must also examine whether and how black boxes, once created, are consulted in the wild—and how such processes may create their own biases.

I focus here on two key stages of the law enforcement officer's interaction with a trucker. First, I consider the interplay between digital enforcement and discretion that arises *before* an inspection occurs, when an officer is determining which trucks to pull in for inspection, and how truckers make use of strategic signaling to evade these checks. Second, I investigate the interaction between digital enforcement and the performance of expertise by considering how the relationship between driver and officer changes when electronic logs are in use *during* an inspection—and how this process can bring about a shift from an adversarial relationship toward one of mutual alliance against the technology.

One note is in order on the timing of our inquiry. Those who study technology face the inevitable hurdle that the technology changes faster than we can write about it. My focus here is on law enforcement interactions with truckers *before* electronic logging was made mandatory—when both paper and electronic logs were legal methods of trucker timekeeping, and when officers were responsible for reading both types. Post-mandate, of course, the situation has changed (in some ways I will highlight). Many of the specific technical and regulatory problems I'll mention have evolved over time and will continue to do so.

But this doesn't mean the lessons of this inquiry are moot; if anything, we learn more about the world during its moments of transition than during its more stable states. Despite the fact that the specific nature of the issues is likely to change, examining the enforcement apparatus at the moment of transition is worthwhile.[4] Doing so makes salient the ways in which interactional norms about power and authority are entrenched in everyday enforcement interactions; viewing the system in an unsettled, interim state—when these norms have been violated—makes it easier to understand what the norms were in the first place. Further, even when the problems people face in managing their relationships around a technology are specific to a particular moment in that technology's integration into a social system, social factors like discretion and interactional dynamics remain perennially important in assessing the workings of the enforcement regime, regardless of its technical specifications.

The Log Inspection Process

Even when a trucker uses an electronic log, the data collected by the ELD are typically read manually by human inspectors.[5] A trucker who exceeds his allowable hours is not *automatically* ticketed by law enforcement, much as we are not (usually) automatically fined for exceeding the speed limit; whether or not we are issued a citation still typically depends, in most cases, on an interaction with a human being. The Department of Transportation has engaged in some pilot projects that would detect hours-of-service violations via an in-road transponder, similar to EZ-Pass Express lanes in which wireless signals transmit data without a car needing to pull over or slow down significantly.

Over the coming years, ELDs are likely to remain human-inspected in one form or another, in significant part because truck inspection processes typically also involve visual inspection of issues like how well a load is secured, whether tire treads are deep enough, and other safety issues that are currently hard to translate into electronic data. Interestingly, law enforcement officers I spoke to have significant trepidation about increasing automation in inspection systems. Officers told me that many significant trafficking busts (human- and drug-related) occur when an officer detects "something hinky" in the course of a routine inspection, possibly as subtle as a trucker appearing unduly nervous or physically sweating during an inspection. Officers worry that if they do not have regular opportunities for close interaction with truckers, they will miss these signals. In any case, a good deal of inspectors' work will still need to be done manually, as some important elements of truck inspection rely on visual and behavioral cues rather than only data that can be transmitted electronically. Therefore, we can expect human inspectors to be part of the equation for a long time to come.

During a log inspection, the officer reviews the driver's logs to ensure that he has not driven more than he is legally permitted to in the past seven days.

Because log falsification is rampant, the officer pays particular attention to the logic of the logs to see if the travel recorded by the driver is sensible—for instance, could a driver realistically have moved from city A to city B in the number of hours denoted on the log? He uses a driver's "supporting documents" (bills of lading, receipts for fuel and tolls) to confirm locations and times when possible.

The inspector also ensures that the "form and manner" of recordkeeping complies with federal regulations—that is, that records are properly kept and signed in required places, that locations where duty status changes occur are noted, and the like. Should any aspects of a driver's log—paper or electronic—fail to meet legally required standards, the officer is required to note the violations on an inspection report. For some violations, the officer can also issue a citation, which is usually accompanied by a fine of hundreds to thousands of dollars depending on the violation—and/or an "out of service" period of two to ten hours, during which the driver may not move the vehicle away from the inspection site. (The logic is that a tired driver should use this out-of-service time to sleep in the sleeper berth of the truck cab.)

All truck drivers are required to pull through a weigh station, where inspections occur, if it is open (usually denoted by a sign or flashing light on the highway). But it would be logistically impossible for officers to inspect the logs of every truck that pulls in. Importantly, the inspector has broad discretion as to which trucks to wave through and which to inspect. To decide whom to inspect, inspectors might look for visual indicators of noncompliance: for instance, inspectors might pull a truck in if it has visible non-log-related violations, like load securement issues, obviously faulty tires, or lack of required permits displayed. Some inspectors told me they use other rules of thumb (pull in every sixth truck) or other strategies. Once performing an inspection, officers continue to exercise discretion: they have latitude to decide how thoroughly to inspect a particular truck, or whether to issue a citation for any violations they find. Truckers may be able to improve their chances by having documents and equipment in good order, and by cooperating with the officer. In the words of one eleven-year veteran officer, attitude counts: "[once] I've looked in your cab, checked your documents, and seen how you handle yourself and how you answer my questions . . . I'm going to make my decision: throw you back in the pond or reel you in."[6] That officer advises truckers selected for inspection to "grit your teeth and go through it with some grace" to reduce their chances of being cited for a violation.

Before Inspection: Strategic Signaling and Decoy Compliance

Electronic logbooks challenge long-held practices in the law enforcement community. Law enforcement officers are used to reading paper logs, which for decades have had the standardized "grid-graph" interface described in chapter 3. Officers have standard protocols for inspecting these logs, involving the order in which they ask questions, the supplemental tools they use during inspection (like mapping software on their own laptop computers to confirm distances between cities), and the methods they use for interrogating drivers about their logs. The grid graph can initially be difficult to decipher for the inexperienced—but over time, officers develop an "eye" for the system and can spot many inconsistencies and violations (at least, those that are visible on the face of the log; as we have discussed, paper logs are readily falsifiable).

Electronic logs, on the other hand, can be much more difficult for inspectors to negotiate for a number of reasons, as we shall see. A key issue during the transition period—that is, when ELDs *could* be used for timekeeping, but were not yet mandatory for all truckers—is that the interfaces electronic logs display for inspection were not standardized. Dozens of ELD models were on the market, and they differed in significant ways. Some ELDs displayed information in a traditional grid-graph format; others listed duty changes chronologically, as a list. Some summarized information like a driver's weekly and daily countdown clock on a separate screen; some listed it on the "home" screen. Some changed the color of certain data to red when the driver was in violation of the time limits (or in danger of having a violation soon); others did not. Some listed time elapsed, rather than time remaining, on a given clock. All the interfaces worked differently— different elements were clickable; some required the use of a stylus; others rendered no electronic display at all, but needed to be printed (some have printers attached). (Eventually, federal rules changed to require ELDs to have a graph-grid display, in large part because "inspectors are used to seeing" information displayed that way.[7])

All of these variations made ELDs much more difficult for law enforcement officers to negotiate, and thus unpopular with a number of them. Some law enforcement officers told me that they were not confident inspecting electronic logs; one noted that some officers were "spooked" by them. In August 2018 (eight months *after* the ELD mandate went into effect), the FMCSA's Enforcement Director, Joe DeLorenzo, told a roomful of truckers that ELD inspections were "equally stressful for both sides"—driver and inspector—and that "the more the driver knows" about the ELD, "the better off that inspection will go"—comments that made plain the fact that inspectors often lacked knowledge about how the systems worked.[8] By no means do I imply that all (or even most) officers had difficulty reading electronic logs; yet this difficulty came up repeatedly in my discussions—and was a common enough belief among truckers to lead them to behave strategically in response to it.

An alternative explanation for inspectors' reluctance to pull in drivers using electronic logs during this period is that drivers running electronic logs were, at least anecdotally, less likely to have violations to find—in part because electronic logs are more difficult to falsify, and in part because the carriers who were using ELDs voluntarily (before the federal mandate) tended to be large, safety-conscious companies, rather than small fly-by-night carriers that were more likely to pose serious safety threats. This would suggest that it would be more efficient for inspectors to spend their time inspecting non-ELD users rather than ELD users. In practice, both dynamics—efficiency concerns and hesitation to engage with the ELD—were likely at work; there are thousands of commercial vehicle inspection officers, and their practices vary widely.

Truck drivers, for their part, were well aware of inspectors' reluctance to negotiate the new technology. Several drivers told me that they were less likely to be inspected when they were using electronic logs. In some cases, drivers reported being pulled into a bay for inspection and then waived through when the officer saw an ELD. As one driver put it: "I've had people like at the coop, the scale house or whatever, where I've pulled in and they ask for my logs and I tell them I've got a Qualcomm [a brand of ELD] and they tell me just to leave." In other cases, drivers reported that they could avoid being pulled in *altogether* by indicating to inspectors that electronic logs were in use in their truck. But how exactly would they do this?

Recall that inspectors pull a truck into an inspection bay from some distance, and often based on visual indicators (insecure loads, flat tires). Knowing this, truckers needed a way to *signal* their use of an ELD from afar. The answer, for many, was the use of a decal affixed to the truck that said as much.

Decals of this sort are typically packaged with ELD systems when they are purchased from vendors (much as a home security system comes with a yard sign or window cling) and state that an electronic monitoring system is in use in the vehicle. A selection of such decals, which I photographed in the course of my fieldwork, is pictured in figure 5.1.

FIGURE 5.1. Electronic log truckside decals. Fieldwork photos.

Because the presence of decals may avert inspection by law enforcement officers, thanks to their degree of discretion as to who to inspect, some truckers and carriers sought to *simulate* compliance with the timekeeping rules by indicating that such a system was in use—even when it was *not*, in fact, in use! Display of a decal without an actual electronic log system to back it up is not against federal regulations—and might allow a driver to "roll through" inspection points. As one trucker put it: "give officers plenty of opportunity to notice you've got an electronic log on your truck . . . and they aren't going to bother you."[9] Being rolled through an inspection saves truckers the possibility of receiving citations if they are in violation of the timekeeping rules, of course—but even when they are "running clean," truckers are interested in avoiding the (unpaid!) delay that inspections impose on their workday.

Because of this, a black market of sorts arose in which some carriers and drivers attempted to obtain decals *without actually installing electronic monitors*. One leading ELD vendor reported to me that, while the company distributed the decals for free, they had restricted the ability to request decals to an existing-customers-only website in recognition of this issue. Another

Qualcomm elog decal source

Co driver and I desire to affix one of those "driver is using electronic logs" decals on our truck. I think it also has the Qualcomm logo.

I searched the web to no avail. Does anyone know of a source? I'll pay postage if anyone has a couple laying around.

My next idea is to check with service centers, but seems silly to waste fuel to acquire a couple decals.

Thanks!

Re: Qualcomm elog decal source

A sign shop can make one up for you. I'm surprised drivers who do not have electronic logs have not yet done so in large numbers. It is not a crime to display such information and the scale cop may roll you through.

FIGURE 5.2. Posts from trucker online forums.

company's sales representative reported that when he traveled to trucking companies to pitch the ELD system, he would complete his presentation only to learn that the company was only interested in obtaining the decals. Truckers have also come up with other creative solutions, such as printing up their own unofficial decals using sign shops or adhesive letters—or seeking extra decals from one another on trucker message boards, as pictured in figure 5.2.

This practice resulted in a form of *decoy compliance*—creating a "false front" that suggests one is following the rules, even when one might not be doing so. This strategy cleverly plays digital and human aspects of the law enforcement regime against each other. Truckers exploited human officers' discretion about whom to inspect, as well as their reluctance to deal with electronic logs, in order to evade enforcement altogether—certainly not the regulatory intent behind the introduction of the electronic monitoring scheme.

During Inspection: Authority through Embodied Performance

Of course, truckside decals weren't a *complete* failsafe against inspection, and inspections involving electronic logs still occurred with regularity. Some inspectors paid no attention to truckside decals; some trucks with ELDs lacked decals; and some inspections occurred for other reasons, like visible violations. Even when inspections of ELDs *did* occur, there were important differences between inspection processes when electronic versus paper logs were in use.

An officer inspecting a paper log typically took a number of defined steps. They would first approach the truck's driver's cab; request that the driver pass the logbook down from the open window or door; ask the driver for additional documentation that is part of the inspection process (load paperwork, permits, a medical inspection card); and then depart for their own vehicle (parked nearby), or the weigh station scale house, for thorough inspection of the drivers' paperwork. There, the officer has a laptop, printer, and other equipment to use in conducting the inspection. Like a police officer pulling over a motor vehicle driver at roadside, a commercial vehicle inspector would run the truck driver's paperwork against a number of electronic databases, state and federal, to check the driver's record, see if the truck is stolen, ensure the carrier has operating authority in the state, and other functions. The officer can also peruse the drivers' logs in great detail. For example, the officer—at their own workstation—could spread out supporting documents (like bills of lading and receipts) to correlate them with stops detailed on the paper log, compare a driver's reported mileage against software that estimates mileages between cities, and use a calculator to add up the multiple rows of numbers that must be checked to ensure the driver's compliance with all aspects of the hours-of-service rules.

Such a process commonly takes as long as thirty or forty minutes, and in extenuating circumstances could take as long as two hours. (These estimates are based on my own observations and on conversations with law enforcement personnel and truck drivers.) During this time, the driver was typically sitting and waiting in his truck cab, unable to depart until the officer finished the inspection. In my observations, officers did not appear to feel any pressure to hurry this segment of the inspection; the process was time-intensive, but the time an officer made a driver wait was an enactment of the officer's authority over the driver. The delay is the officer's prerogative. And because the work is detail-oriented, it cannot be rushed.

The physical separation of driver from log also facilitates interrogation of the log's contents. An officer would often ask the driver to pass down his log, and only *then* would they ask questions of the driver ("Where did you start this morning?" "What time did you finish driving last night?") that were intended to assess the veracity of the entries on the log. The separation of driver from log serves an important function: it is intended to keep the driver from "cheating" on those questions by giving the answers detailed on the log.

But this process differed fundamentally when an officer inspected electronic logs. First, because of the variety of interfaces being used at the time, officers often had significant difficulty acquiring time data from the monitors. Because the data were presented in a wide variety of formats, officers generally lacked fluency with all the possible systems with which they might be presented. Foreseeing this difficulty, the federal regulations pertaining to ELDs required drivers using electronic logs to carry an instruction card that must be tendered to the officer detailing how data can be retrieved from it.[10] Though failure to carry such an instruction sheet is an inspection violation, many drivers lacked the card, or confused it with other documentation related to the ELD provided to them by their carrier. Even when such a card *was* provided, an officer's reliance on it made abundantly clear to the driver that the officer lacked inherent knowledge about how to operate the equipment, greatly undermining the officer's authority.

During the time of my fieldwork, physical separation was typically not an option when electronic logs were in use.[11] Because electronic monitors were hardwired into the trucks, they could not be separated and taken back to the officer's vehicle for close, isolated inspection. Some models were equipped with cables that allow the monitor to be stretched out the truck window, meaning that the officer must stand outside the cab on the ground, possibly in inclement weather, and below the (seated) driver, to complete his log inspection. But most models lacked a cable, or at least one long enough, and had to be inspected on the dashboard where they were attached. This required the officer to enter the truck's cab and complete the inspection there, often sitting in the driver's seat of the truck, with the truck driver standing behind him, quite literally looking over his shoulder as the officer pokes at the ELD screen. Truck cabs are not large spaces, and they function as truckers' homes when they are on the road. They may be quite dirty or crowded with personal belongings. Some truckers bring dogs with them on the road, which may be resident in the cab during the inspection. And, of course, entering this small, unfamiliar, intimate space requires the officer to attenuate their attention from the ELD in order to keep an eye out for their own personal safety (for instance, as one officer reported to me, he was careful to position one elbow over his firearm holster while seated in the cab during an ELD inspection, and did his best to survey the cab both upon entry and continually for weapons or contraband).

Thus, when electronic logs were involved, inspections were more difficult for officers to complete thoroughly, for both technical and interactional reasons. This meant that logs were inspected *less carefully*, and that inspections *took much less time* (often five to ten minutes). Of course, this difference could be interpreted as electronic logs having precisely the effect they are designed to: there is no reason for officers to complete time-consuming, extensive interrogations of hours-of-service records if the electronic system essentially assumes this burden for them. But attention to the interaction between enforcer and enforcee suggests that this difference occurred not *only* because electronic logs may have been inherently more accurate than paper logs, but perhaps also because they were more technically difficult and socially uncomfortable for officers to negotiate.

The Digital Common Enemy

One other interesting dynamic emerged in these interactions. When officers inspected electronic logs, the interaction between enforcer and enforcee could shift from being an adversarial one—aimed at uncovering inconsistencies in accounts and probing the trucker's veracity—to being reluctantly collaborative. The officer depended on the driver for help in retrieving data for inspection; as described, this dependence is even formally inscribed in the law, since an instruction card must be tendered by the driver. Any difficulties or incompetencies the officer displayed in dealing with the ELD occurred in full view of the driver. This could create a counterintuitive and almost poetic inversion in which the trucker—who might normally bristle under the watchful eye of law enforcement—instead becomes the observer, quite literally looming over the shoulder of the officer.

In some cases, the oppositional relationship between driver and officer could morph into one of *new alignment* against the technology, which becomes the central point of focus in the interaction. A driver might lack (or, at least, claim to lack) knowledge about the workings of the monitor, saying something like "oh, the company made us put these in"—and drivers' complaints about the ELD might land on surprisingly sympathetic ears. Because most drivers did not choose to install the monitor and are not the direct creators of the time entries it displays (as compared to the pencil markings created by the driver on a paper log), it is easier for the driver to disclaim responsibility for the data the ELD renders, and to be at least partly absolved of it ("I don't know why it says Colorado Springs! I was in Denver!"); after all, who among us has not had our efforts thwarted by technology on occasion?

In some cases, inspectors might respond with commiseration—sometimes subtly, via a shake of the head or a meaningful sigh—or by calling attention to the poor design of the systems. As one trucker recounted in an *Overdrive* magazine feature on ELD logging errors, the officer was "'in the truck with me, kind of agreeing,' nodding along with [the trucker's] tour through the device," showing some sympathy for the trucker's plight. (He eventually issued him a citation anyway.)[12] In other cases, inspectors unable to parse ELD data redirected their enforcement attentions to more cut-and-dry requirements, like the presence or absence of the inspection card, instead of putting their expertise on the line by attempting to closely interpret the monitor.

The ELD could serve as almost a third party in the enforcement interaction: a convenient common enemy for the driver and the officer, against which they may be temporarily united in frustration and lack of understanding. In this way, the technology can occasion momentary solidarity, transforming an adversarial relationship into a temporarily aligned one, as it provides an avenue through which the inspector can save face[13] and regain authority that he or she may have lost due to lack of demonstrated expertise with the technology.

This uneasy alliance could extend beyond the immediate inspection interaction, too. Some law enforcement agencies went so far as to express their skepticism about ELDs directly to truckers on their social media pages. Just before the mandate, the Indiana State Police's Commercial Vehicle Enforcement Division invited truckers to post memes making fun of ELDs, saying "I think we can say that WE are all flustered with the ELD mandate . . . so please post your best ELD meme below!!! Keep it clean but let's see them and share a laugh!!"[14] Hundreds of truckers posted anti-ELD comments and jokes, many of which the police responded to with good humor.

Surveillance, Performance, and the Destabilization of Power

The difference between interactional dynamics in paper-based versus electronic inspections underlines the embodied nature of power and authority in the enforcer-enforcee relation. When a paper log is used, time and distance are sources of authority for officers—who strategically separate the driver from the log, peruse the logs carefully in their own spaces with their own tools, and make drivers wait as long as the process requires. When electronic logs are used, officers cannot draw upon these resources; their authority and expertise are on the line as they attempt to complete their duties in full view of drivers.

These interactional differences are not captured in the common rhetoric of technology as a perfector of enforcement. Carriers and drivers exploit officers' perceived reluctance to deal with electronic monitoring systems (whether as a result of officers' lack of expertise or their assumption of legality) by signaling compliance using truckside decals. This practice of "decoy compliance" underscores the central role of discretion in the enforcement regime—despite the assumption that digital enforcement *minimizes* discretion. When inspections of electronic systems do occur, the traditional power relation is destabilized as the officer's authority is undermined and new vulnerabilities are created, due to the visibility of the inspector's (lack of) expertise; in order to "fix" this encounter, officers and truckers may align temporarily against it.

Most of my focus in this book has been on how the ELD controls truckers. But commercial vehicle inspectors, too, are disciplined by this technology with which they are tasked to work. The labor of sensemaking in the inspection process is far from trivial. The regulations to be enforced are dense and complicated; officers must simultaneously look for rule violations, corroborate multiple sources of evidence to root out inconsistencies and likely falsehoods, remain vigilant about their own safety, and keep alert for signs of more serious wrongdoing, like human or contraband trafficking. When an inspector's cultivated routines for conducting this sensemaking work are disrupted by new technologies—putting their newfound lack of skill on full visual display in front of the person over whom they are assumed to have authority—the officer may lean on their inherent discretion to avoid the encounter altogether; or, they may conduct their work in a much more cursory way, or take steps to "fix" the encounter through commiseration with the driver.

Understanding these dynamics also forces us to reevaluate the measurement of the effects of surveillance tools. The reduction in recorded timekeeping violations under electronic monitoring might mean that drivers are breaking the rules less often, leading to less need and opportunity for citation; this is certainly the explanation preferred by policymakers who espoused the safety benefits of the ELD (however, recall from chapter 3 that evidence seems to suggest that drivers compensate for lost flexibility in other ways, such as speeding). But an alternative explanation exists—if surveillance technology (or the perception of surveillance technology!) drives *different enforcement activity* by inspectors, as it indeed seems to do, *that* may be driving the changed numbers instead. Theoretically, this effect could occur even if truckers didn't adjust their behavior at all under electronic monitoring. During the pre-mandate period, small carriers (far less likely to use ELDs) accounted for a much larger proportion of hours-of-service violations than large carriers did—but they were also inspected much more often, with *more than twice as many* reported inspections per truck for carriers with one to four trucks versus 500+ truck carriers.[15] (And this doesn't even account for the more cursory treatment that large carriers, more likely to have ELDs, might have received even when they were inspected.) What looks like an improvement in safety stemming from ELDs might actually reflect how much inspectors are looking for violations when ELDs are in use.

Conventionally, we tend to think of surveillance technologies as reinforcing traditional authority relations: states and other powerful actors use these tools to reassert and deepen their authority over surveilled subjects. The case of the ELD complicates this understanding by suggesting that the small-scale interactional practices around these technologies matter, and in some cases can rearrange observer/observee relations in ways that surprisingly *de*stabilize power dynamics.

6

Beating the Box

HOW TRUCKERS RESIST
BEING MONITORED

The ELD is intended to compel truckers' compliance with the rules by quashing their long-held ability to "fudge" paper recordkeeping. As we've described, noncompliance of this sort is culturally encouraged, economically incentivized, and technically easy; it involves only paper, pencil, and (if asked to explain oneself) the ability to justify one's story with a straight face. But as ELDs have pervaded the industry, noncompliance strategies have evolved, too. Truckers are a savvy and motivated bunch, with a deeply autonomous occupational culture—and no digital system is completely foolproof. In this chapter, I explore how truckers, firms, and other key players "beat the box" by adopting a variety of strategies of resistance against monitoring.

Resistance is not only important for truckers: it plays an important role in how people respond to all kinds of monitoring in all kinds of circumstances. As surveillance continues to become more pervasive across daily life, and as it becomes less feasible to "opt out" wholesale, people adopt new and creative strategies that challenge, obfuscate, and thwart data collection. Sometimes, popular rhetoric about surveillance tends to render individuals as merely passive data points, being manipulated by top-down systems without the ability to respond. But a focus on resistance recognizes the agency individuals have in the face of these systems. Surveilled parties are more than simply "docile bodies" (as Foucault characterized them)[1]—they're actively engaged in surveillance regimes in many ways, and often work to foil them.

Scholars of surveillance have taken note of these practices. Perhaps the most systematic scholarly effort to categorize resistance strategies to date is Gary Marx's 2003 article "A Tack in the Shoe"—its title a reference to the notion that stepping on a tack hidden in one's shoe can help one beat a polygraph test, by distorting the physiological indicators which are meant to indicate veracity or falsehood.[2] Marx catalogues many different sorts of "moves" people engage in to subvert and resist surveillant systems, from avoidance (like using an untraceable phone number) to piggybacking (illicitly entering a doorway immediately behind a person with an access card). Finn Brunton and Helen Nissenbaum develop the idea of *obfuscation* as resistance practice—the practice of producing false or ambiguous data to confuse a surveillor, or to make surveillant practices more costly.[3] Other scholars suggest that *sousveillance*—inverse surveillance, or "watching the watchers"—can be an effective countermeasure.[4] Torin Monahan's study of counterstrategies aimed at urban closed-circuit television systems identifies a range of practices in which people engage—from destruction to evasion to artistic protest—but suggests that these tactics often fail to challenge the deeply embedded institutional basis of surveillance in lasting ways.[5] Other influential studies examine strategies of political resistance: most notably, James C. Scott's *Weapons of the Weak*[6] illustrates the quotidian tools of defiance engaged by the powerless against the powerful—among them foot-dragging, gossip, and sabotage—and how these small moves help people push back on the regimes that oppress them. This rich line of research informs our examination of truckers and how they push back against electronic monitoring.

As I have described previously, there is a strong cultural emphasis on freedom, machismo, and independence in the trucking workforce. Truckers are often loath to be told how to do their jobs, and are often defiant or derisive in the face of government or organizational bureaucracy. As such, they are an ideal population to study in order to learn about methods of, and motivations for, resistance to surveillance and control. Identity is central in resistance practices: resistance to surveillance is most likely to occur when the identity of the subject presumed by the surveillant system is fundamentally incompatible with the surveilled party's understanding of themselves.[7] This dynamic is acute in the truckers' case. Truckers resist surveillance to preserve their coveted privacy and independence, and to reestablish their long-held occupational autonomy. To the extent that drivers circumvent surveillance to make more money, that motivation may be supported by a strong cultural norm toward provision for one's family. And the idea of being able to withstand exploitative and incredibly difficult work conditions accords tightly with truckers' sense of professional pride. Resistance, then, is a means of asserting one's professional identity as a trucker, for better or worse. Being electronically tracked, on the other hand, is completely antithetical to truckers' professional identity.

But as we'll see—the story is not quite this simple. To be sure, some of truckers' resistance *is* driven by truckers themselves, as they seek to reassert their identities, to preserve their occupational autonomy, or, quite simply, to do what needs to be done to get the job done and make a living. But some forms of resistance against the ELD are more complicated, involve more people, and—crucially—sometimes work *against* truckers' interests. A key emphasis of my analysis here is to consider *what purposes resistance serves, and for whom*. We'll see that resistance ends up being a process through which truckers, companies, law enforcement, and others negotiate power among themselves, and often to different ends than we might initially have anticipated.

Truckers' Resistance

Truckers employ a panoply of resistant maneuvers to reassert their autonomy against electronic monitoring—ranging from physical tampering to data manipulation, from technical workarounds to political action. Some strategies involve collusion among drivers and their employers, or seemingly disinterested third parties, like gas stations. Some tactics rely on coordination or information-sharing among drivers, while others are more independent in nature. Resistance practices can serve multiple purposes for drivers, from expressive interests to instrumental economic benefit.

I uncovered a dozen resistance strategies truckers use or have used, which I place in four thematic buckets:

- physical tampering with the monitoring system
- manipulation of data captured by the system
- exploiting the technical limitations of the system
- social and organizational resistance *around* the system

In examining how truckers resist, I looked at both new and old strategies—those that emerged anew with the rollout of the modern ELD, and those that have adapted over time in response to changes in timekeeping technology. Resistance to surveillance is difficult to study empirically: after all, its success often (though not always) depends on being covert in order to avoid detection by surveillors. In Appendix A, I describe in more detail some of my research approaches for uncovering these strategies.

It's important to note that many of the strategies I describe have "gone stale" and might no longer work as well as they initially did. Technology vendors and truckers are locked in an arms race of sorts, as ELD manufacturers figure out ways to close loopholes as they are exploited. But no-longer-effective strategies are just as informative as those that still work; the goal of this chapter, after all, isn't to serve as a manual for beating the ELD, but instead to learn from truckers' responses to it.

Reporting on no-longer-effective strategies also has some ethical advantages. During the course of my research, I was sometimes contacted by trucking technology firms who were keenly interested in my work on ELD resistance and would invite me to come share my findings with them. It was clear from the nature of these invitations that these firms weren't interested in my study for social science's sake; rather, they wanted to know more about how their systems were being tampered with so that they might design their tools to prevent such resistance in the future. I never accepted these invitations, as I didn't see my role as serving as "tattletale" to these firms. Reporting on stale strategies helps to alleviate some of these concerns—the reason the strategies no longer work is often because firms already know about them.

Physical Tampering

Some forms of resistance involve physically damaging the surveillant system to impede its physical function. Truckers used different techniques to destroy the monitoring devices in their trucks—some quite overt, and others more subtle.

BRUTE FORCE

Perhaps the most direct way to resist a surveillant system is, quite literally, to smash it. Some truckers employed "brute force" methods of resistance by taking a hammer to the offending computer.

Brute-force destruction can be an overt way for a trucker to symbolically communicate with his (perhaps soon-to-be-former!) employer—signaling in no uncertain terms his feelings about being monitored. In other cases, the driver might take steps to preserve the outward physical appearance of the monitor in order to destroy its capabilities more covertly (and deniably). One industry association reported that drivers destroyed monitors by "covering them with a small bag of dry ice and tapping them with a rubber mallet." Doing so left "no outward sign of assault"—but left the machine's "solid-state innards shattered into a billion pieces" and unable to function.

Brute-force destruction can also be a means of coordinated, collective dissent among multiple drivers. One former driver told me that *all* the drivers at his furniture-delivery firm engaged in coordinated brute-force resistance when the company installed timing equipment in all of its trucks. Upon leaving the warehouse, he reported, "the driver would wrap his jacket around the timing instrument and disable it by hitting it with a hammer. All of the drivers participated, and this continued for about a week, at which time the company gave up on its timing attempts, assuming, I suppose, that the timing instruments just did not work."

A popular trucker video on YouTube makes light of the situation. The video begins with a trucker pressing a variety of buttons on his ELD and entering seemingly random combinations of digits in various fields, while promising: "I'm gonna show you how to cheat on the e-log system. How to hack it! How to get rid of it!" The video then abruptly cuts to an image of him smashing the device with a sledgehammer: "So basically what you do is, you take your tablet and you take this wonderful hammer and you just beat the crap out of it. And what do you know, after a couple good strong hits, you can break your tablet in half and go back on paper. No more e-logs!"[8]

Truckers aren't the only ones who sabotage their monitors. In 2016, an investigation of Chicago Police Department dashcam videos found that *80 percent* of captured videos were missing audio due to officer error and intentional damage to the equipment. Officers had removed batteries, busted antennas, and thrown parts away.[9] Similar problems plagued equipment at the Los Angeles Police Department; sociologist Sarah Brayne describes a "rash of antennae malfunction" caused by officers deliberately removing antennae to prevent higher-ups from being able to hear what they were saying.[10]

To be sure, brute-force techniques can be risky (and expensive!). But there's a certain clarity to taking out one's frustrations on surveillance with a hammer. It's a visceral response to an unwelcome intruder in the truck cab.

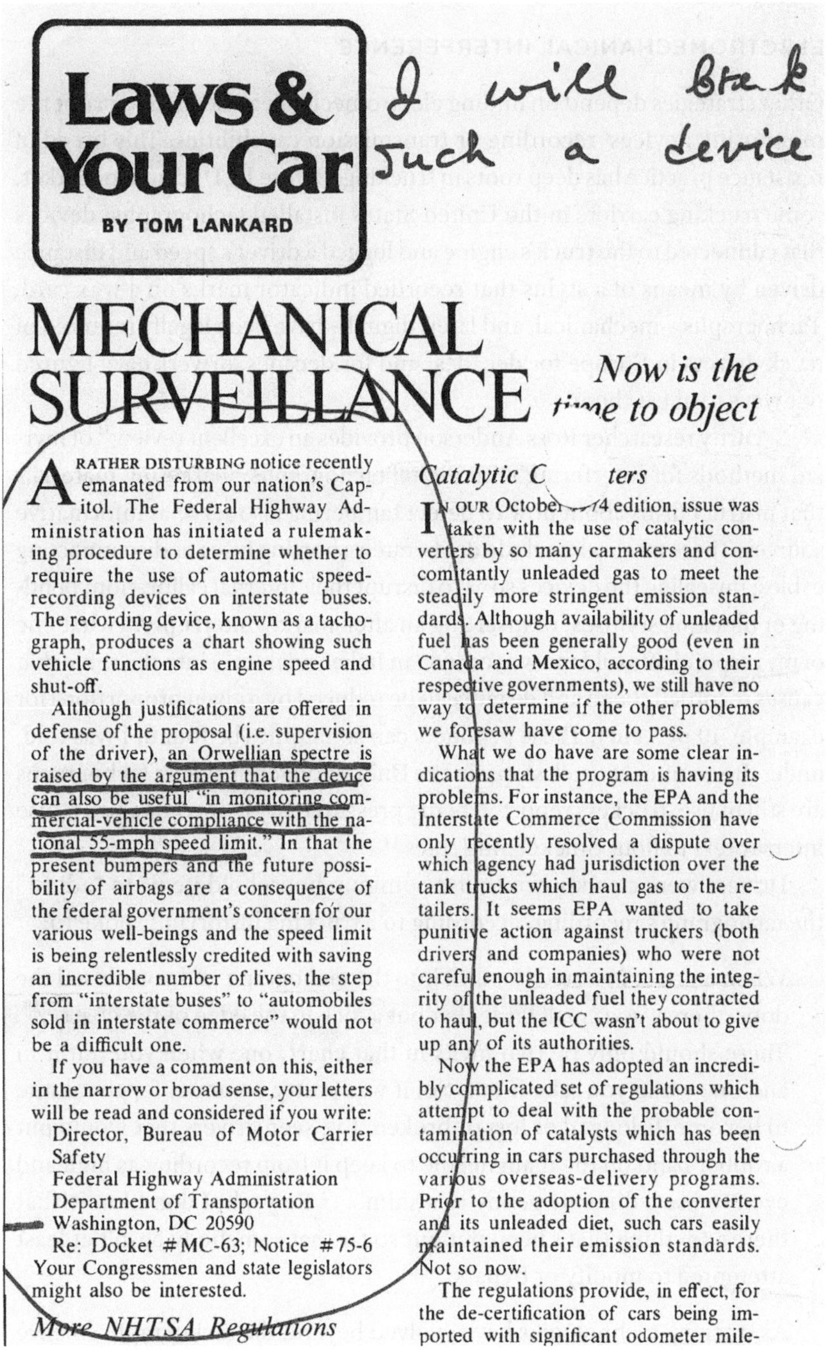

Laws & Your Car

I will break such a device

BY TOM LANKARD

MECHANICAL SURVEILLANCE

Now is the time to object

A RATHER DISTURBING notice recently emanated from our nation's Capitol. The Federal Highway Administration has initiated a rulemaking procedure to determine whether to require the use of automatic speed-recording devices on interstate buses. The recording device, known as a tachograph, produces a chart showing such vehicle functions as engine speed and shut-off.

Although justifications are offered in defense of the proposal (i.e. supervision of the driver), an Orwellian spectre is raised by the argument that the device can also be useful "in monitoring commercial-vehicle compliance with the national 55-mph speed limit." In that the present bumpers and the future possibility of airbags are a consequence of the federal government's concern for our various well-beings and the speed limit is being relentlessly credited with saving an incredible number of lives, the step from "interstate buses" to "automobiles sold in interstate commerce" would not be a difficult one.

If you have a comment on this, either in the narrow or broad sense, your letters will be read and considered if you write:

Director, Bureau of Motor Carrier Safety
Federal Highway Administration
Department of Transportation
Washington, DC 20590
Re: Docket #MC-63; Notice #75-6

Your Congressmen and state legislators might also be interested.

More NHTSA Regulations

Catalytic Converters

IN OUR October '74 edition, issue was taken with the use of catalytic converters by so many carmakers and concomitantly unleaded gas to meet the steadily more stringent emission standards. Although availability of unleaded fuel has been generally good (even in Canada and Mexico, according to their respective governments), we still have no way to determine if the other problems we foresaw have come to pass.

What we do have are some clear indications that the program is having its problems. For instance, the EPA and the Interstate Commerce Commission have only recently resolved a dispute over which agency had jurisdiction over the tank trucks which haul gas to the retailers. It seems EPA wanted to take punitive action against truckers (both drivers and companies) who were not careful enough in maintaining the integrity of the unleaded fuel they contracted to haul, but the ICC wasn't about to give up any of its authorities.

Now the EPA has adopted an incredibly complicated set of regulations which attempt to deal with the probable contamination of catalysts which has been occurring in cars purchased through the various overseas-delivery programs. Prior to the adoption of the converter and its unleaded diet, such cars easily maintained their emission standards. Not so now.

The regulations provide, in effect, for the de-certification of cars being imported with significant odometer mile-

FIGURE 6.1. A document submitted to the Federal Highway Administration in response to a 1975 proposal to require mechanical tachographs in some commercial vehicles: "I will break such a device."

ELECTROMECHANICAL INTERFERENCE

Other strategies depend on finding electromechanical means to disrupt the monitoring devices' recording or transmission capabilities. This breed of resistance practice has deep roots in trucking. Before ELDs came to market, some trucking carriers in the United States installed tachographs: devices that connected to the truck's engine and logged a driver's speed and distance driven by means of a stylus that recorded indicator marks on a wax card. Tachographs—mechanical, and later, digital—have been legally required of truck drivers in Europe for decades; and for decades, drivers have figured out ways to cheat them.

Security researcher Ross Anderson provides an excellent review[11] of myriad methods for interfering with digital tachographs; marketing materials that instruct firms about how to detect tampering also serve as informative sources. These methods include deliberately shorting fuses or disconnecting cables; unsealing the devices so as to disrupt their internal calibration; bending or blocking styluses; or inserting an after-market interrupter (what one of my respondents told me was called "an Italian adapter") into the cable that causes recorded speed and distance to be reduced by a given proportion (for example, 10 percent). These products can sometimes be found, I was told, under the counter at truck stops. In the United Kingdom, where tachographs are still in use, truckers reported being pressured by their companies to use interrupters to flout timekeeping rules.[12]

Drivers were also known to use common household items to "adjust" the tachograph's recording, according to a trucking historian I spoke to:

> When a driver inserted the disk into the tachograph and you closed the door, there was a small knife that put a nick in the edge of the chart. . . . There should only be two nicks in that chart, one when you put it in and one when you take it out. So, it was pretty common for that knife to become dislodged or lost or broken. I've seen drivers that would put a rubber band on the main needle to keep it from recording as high and even to use a dime or penny as a shim. . . . So yeah, I don't know that there's anything that's been put out so far that somebody hasn't at least attempted to modify or bypass.

As tracking technologies have evolved beyond the tachograph, so have schemes for tampering with them. Most devices use GPS to track a truck's location; some drivers cover the GPS transmitter on the truck's roof with tinfoil or a large metal bowl, which can prevent a satellite from detecting

Evidence of Tampering

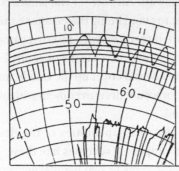

Opening or Closing of the Tachograph Door

The first place to look for tampering in analyzing the charts is the cuts made by the notching blade upon opening and closing the Tachograph Door. Unauthorized opening or closing of the door is an indication to the chart reader to examine that chart more carefully.

In addition, each time the door is opened the styli are not in contact with the chart and no recording is registered. The styli will not return to their original position when the door is closed, and the break in the styli markings will be apparent.

FIGURE 6.2. Tampering detection information from a 1970s tachograph manual. *Source*: "Tachograph Chart Reading: A User's Resource Guide," Gemco Truck Products Division/Engler Instruments Operation.

the transmitter. Message boards are replete with drivers anonymously sharing tips for tampering with GPS receivers and other tracking technologies:

> fyi, there are TWO power wires so if you are thinking of unpluggin [*sic*] from the fuse panel you have to unplug both of them, with one unplugged the keyboard will appear dead but the system can still record certain things, the company you work for can still query the system directly and get a response[.] with both wires unplugged the system is completely shut down, at least the older systems are. one wire is a 5 amp and the other is 25 amp.

> on most systems you can also turn the key on and off several times and it will shut the system down, I do this a lot to conserve on battery power.

> just do as I do and wrap a piece of Aluminum foil around it and secure it with small bungy cords. this will stop the Transmit and receive but, it will not prevent it from recording your speed or how many times you stop.

Other drivers purchase illegal after-market GPS jammers intended to accomplish the same thing. Interestingly, these devices can have unexpected externalities. A 2013 *Economist* article described how truckers' GPS jammers can disrupt timestamps on financial trades at the London Stock Exchange, navigation systems in cars, mobile phone signals, and aviation systems.[13] In 2013, a New Jersey truck driver was fined $32,000 after a $100 GPS jammer—which he had plugged into the truck's cigarette lighter to avoid being tracked by his company—caused interference with a system used to guide pilots to the runway at Newark Airport.[14] (The jammers are also popular with truck thieves in order to avoid being traced after a theft.)

Some fleet management systems are integrated with internal cameras directed at the driver, which record his actions as a means of control and training. Drivers reported to me that they would cover up the small camera lenses with duct tape or popsicle sticks, or would draw over them with black Sharpie, in order to avoid being watched.

Of course, as with most technologies, the capabilities of the device advance over time to thwart common tampering schemes like these. For instance, some devices now record data streams in two parallel encrypted formats; if the two streams do not match, it raises a red flag that the device has been tampered with. And generally, the systems signal irregularities to the company (for example, if a satellite signal is temporarily lost), and vendors often supply companies with training that explains how to detect tampering of this sort.

Manipulating Data

A second set of strategies involves manipulating the digital data ELDs collect in order to violate the rules.

DATA EDITS

In the most overt case of data manipulation, truckers and firms affirmatively misrepresent their driving time by editing ELD data in the "back office." ELD software typically provides the capability for managers to make retrospective edits. There are plenty of situations in which a manager might legitimately need to edit a trucker's data (for instance, when a driver forgets to log in; an ELD malfunctions; the satellite signal cuts out in a rural area and fails to capture a change of status). But some truckers told me that dispatchers sometimes fudge ELD data in order to give drivers extra time behind the wheel, so that they can make a crucial pickup or delivery window without "service failing" the load.

Before ELDs were mandatory, this technique didn't seem to be widely used. If a trucking firm was of the type that would routinely make data edits to facilitate illegal activity, it was also unlikely to have voluntarily adopted electronic monitoring. It's likely that those firms relied on paper logs—which are much easier to falsify—for as long as they could. But we might expect strategies like these to become somewhat more common when monitoring technologies are mandatory.

Perhaps the biggest reason data edits had not come into broader use as a resistance strategy is that ELD software commonly flags data that have been retrospectively edited, leaving a digital paper trail showing something was changed manually. In computer security, access to data can be controlled by a mechanism called "breaking the glass"—a means of manually overriding general access restrictions (for example, to gain access to information to which one would not typically have access) in exigent situations.[15] The system grants access, but also automatically sends a notification to the user's superior of the access. The design aims to preserve the flexibility of the system and recognize the need for occasional override, while creating a measure of accountability to prevent unauthorized use. In a similar move, to manually edit ELD data, the dispatcher or fleet manager often is required to input a text explanation for any manual changes, especially changes involving driving time. These data are also flagged and would be likely to undergo additional scrutiny in the case of an internal or government safety audit. ELD rules also require edit logs to be visible to roadside inspectors—a further check against the utility of this strategy.

"HACKING"

Some drivers developed methods to circumvent restrictions imposed by their companies on ELD use—what they sometimes referred to as "hacking" the device. Though drivers in my research had not (yet!) devised ways to digitally hack into the systems in order to directly falsify data about their driving time, some drivers *had* discovered ways to use ELDs for other unsanctioned activities. For instance, a YouTube video posted by one trucker—which has been viewed over sixteen thousand times—shows how to access a solitaire game on his "built-in snitch," the Qualcomm MCP200. Though this behavior may seem trivial—this guy just wants to play solitaire!—the video makes very clear how meaningful the act is. The trucker's narration of the hack depicts deep resentment of the electronic device and scorn for the company that requires it. In a derisive tone of voice, he says:

Say you want to get to the games, but say that your douchebag company says, "hey, you don't get to play games. We're gonna gray that shit [the interface button for games access] out so you can't fuckin' touch on it no matter what." Here's what you do.[16]

The trucker then demonstrates what you do: he presses the top of the screen seven times, then the control and escape buttons on the keypad, which brings up the Windows XP start menu. From here, he shows how to access built-in games as well as a web browser; he also describes how to install other games, like Quake, using a thumb drive. (Comments on the video indicate that this workaround is no longer effective; commenters presume that subsequent software updates blocked this capability.)

The trucker's action is far from playful. Despite the fact that the circumvention does not get him any additional driving time, it is nonetheless an act of defiance that evinces his contemptuous relationship to the technology and the company that put it there. It's temporary, evasive, and unlikely to significantly upend structures of power—nor is it seemingly intended to do so. But this makes it no less an act of resistance. As Ewick and Silbey write: "The meaning of what seem like petty acts lies in their narratives."[17]

Exploiting Limitations

A third bucket of resistance strategies involves a different sort of toolkit. Rather than manipulating the ELD physically or tampering with its digital data, drivers exploit the *limitations* of the technology. By knowing what the ELD *can't* detect, drivers and firms understand what loopholes they can make use of to eke out extra driving time or avoid following the rules.

GHOST LOGS

Most ELDs require drivers to log into the system using a username and password in order to authenticate themselves. But a driver often has access to multiple username/password combinations—perhaps because another driver has "lent" them to him, or because the company has furnished multiples to him, or because he knows of a demo or dummy account he can use (which many firms have, either for training purposes or to cover lags between employment of a driver and the time at which he's entered into the electronic system). When he's reached his legal driving limit, he can simply log off . . . and then log in again as another user. Though the "new" driver is the same person, the monitor is none the wiser. This practice descends from a similar practice involving paper logbooks, in which drivers would simply keep multiple sets of logs in the truck.

When "ghost logs" (as they are known) are used, they create holes in the journey that must be accounted for. Imagine that driver John's ELD records him driving from Indianapolis to Nashville before his hours run out; John then goes off duty while a ghost driver, Casper, logs in and takes the load from Nashville to Atlanta. In Atlanta, Casper logs out and John logs in again, having appeared to have gotten some restorative, and legally required, time off work. If John's logs are inspected in Tallahassee, a competent inspector will question how he and his truck managed to get from Nashville to Atlanta while he was off duty. John might simply explain that he was "driving team" with Casper from Nashville to Atlanta; that is, he was resting in the sleeper berth while Casper plowed ahead. Where is Casper now? Well, he got out at Atlanta, because the company needed him to pick up a load there. This explanation accords with industry practices (firms do have to move trucks and drivers around the country in various ways to prepare for upcoming hauls, a practice called "power balancing"), and has the additional virtue of being difficult for inspectors to contest.

MIS-LOGGING NON-DRIVING WORK

Though ELDs automatically track driving time and location, they fail to capture other types of data that might more accurately reflect a driver's actual work time. The most significant of these is "detention time"—the time drivers spend waiting to be loaded and unloaded at shippers' and receivers' terminals. Recall from our discussion in chapter 2 that these wait times can be excessively long, and that drivers are usually unpaid for this time, though they often need to remain alert (to supervise loading processes, to move the truck around the terminal, or to safeguard the truck's freight). These wait times can cause significant problems for legal compliance, as a driver facing a long delay may run out of available hours to drive to his next destination (recall that the driver may be on duty for only one fourteen-hour window per day). Many drivers maintain that it is this long, uncompensated work time that creates the *real* fatigue and unsafe working conditions in the industry.

Drivers are legally required to log detention time as "on-duty, not driving"; this time counts toward a driver's fourteen-hour daily window of work time and his seventy-hour weekly limit. But a driver has little incentive to log this time accurately, because exceeding these limits interferes with the time when he can legally drive (and make money). ELDs are *not* capable of detecting work activity when a truck is not moving; thus, loading/unloading time is manually entered by drivers into the devices, and many drivers log it as off-duty rest time instead, in order to preserve work hours. Drivers often similarly mis-log other nondriving work duties, like legally mandated daily inspections of the truck (which, if done thoroughly, can take half an hour). Mis-logging often takes place with the knowledge, and sometimes under the explicit direction, of trucking firms, as we'll see in the next section.

ROLLING TO STOPS

Another method of exploiting the limitations of electronic monitoring involves strategically gaming technical thresholds. Many ELD models categorize a truck as "driving"—and thus, its driver as being on duty—when it is detected as having moved a certain distance, or at a certain rate of speed; these thresholds are built into the devices, though organizations can often set the specific limits via back-office preferences. But in practice, dispatchers can conspire with drivers to avoid triggering these thresholds and to buy some extra work time.

In one of my observations at a trucking firm, a driver had very little time remaining on his fourteen-hour duty and eleven-hour driving clocks, yet was still a few miles from his pickup point. The dispatcher instructed the driver to park the truck at a Walmart he was approaching and to record himself as being off duty on his ELD. (Walmart stores are well-known among truckers for often allowing trucks to park in their massive lots. In fact, given the dearth of public funding for truck parking, many drivers have come to rely on the national network of Walmarts as *de facto* truck stops where they can sleep undisturbed and stock up on supplies.) At that point, the driver was legally required to take a ten-hour rest break in order to gain more work time. But the dispatcher instead instructed the driver to "roll" the remaining handful of miles to the pickup point, *at a speed of less than fifteen miles per hour*, in order to avoid triggering the ELD from registering the truck as "driving." Once the driver reached his destination, the dispatcher instructed him to load the freight—which *also* could not be detected by the monitor. The driver did as he was instructed, rolling and loading, all while remaining (officially) on an unpaid off-duty rest break.

What was especially striking about this incident to me was that the dispatcher had, not an hour before, told me that the carrier for which she worked "ran 100 percent legal." The company was proud of its high safety rating and distanced itself from carriers that racked up violations of the hours-of-service rules. Yet, in full view of an outside observer (me), this dispatcher did not hesitate to tell the driver how to break the law without getting caught. When the dispatcher hung up with the driver, I had fully expected her to look at me sheepishly, or to offer some awkward explanation for why lawbreaking was justified in this instance. But nothing of the sort happened; instead, she continued on with her work, moving on to the next driver interaction without justifying the circumvention to me.

Similar incidents occurred in my observations at other field sites, in which fleet managers would inform me that they were strict about not letting drivers exceed their legal limits, as evidenced by their voluntary adoption of ELDs. But when I watched those managers confer with drivers about (most typically) how to log undetectable nondriving work time, they told them, explicitly or implicitly, to mis-log it. In one instance, for example, a dispatcher told me that the company had determined that a legally compliant truck inspection could take place in eight minutes, so drivers should log that amount regardless of how long the inspection actually occurred—or for that matter, whether it occurred at all.

The practices suggest a conflation of legality with the avoidance of detection. "Officially"—if by official, we mean verifiable according to electronic records—these firms are operating legally. But by exploiting the limitations of the recordkeeping capabilities of the monitors, firms can *appear* compliant while running their operations in ways that are not, in fact, compliant. Dispatchers and fleet managers were quite casual in revealing these practices to me.

The organizational sociologist Diane Vaughan notes that during periods of rapid change, firms may engage in unlawful behavior because "'what is very generally at stake is the definition of what is deviant.'"[18] Here, firms' personnel seem not to consider these practices to constitute deviance—a construction that is unsurprising, given that it aligns with both long-held industry norms and the necessity of operating flexibly under the new technological regime.

USING OLD OR DUMB EQUIPMENT

Technology and the rules that govern it don't evolve at the same speed. Often, technology outpaces the law: it can take years for a policy to make its way through plodding bureaucratic processes, while technological advancements race ahead. This difference in speed is sometimes called the *pacing problem*,[19] and is a key reason why it can be so difficult to make forward-looking technology policy that isn't outdated by the time it takes effect.

But sometimes the law must contend with the opposite problem: some people *don't* have the latest, state-of-the-art technologies, and it would be unrealistic or unfair to expect everyone to adopt them quickly. So while we may want to make policies with the latest technologies in mind, we also have to deal with the lag of what people actually own and use, which may not always fit the rule as well. The law has to deal with both of these realities at once: innovation happens quickly, but it doesn't happen evenly.

A common way to deal with this in law is through including a "grandfather clause" in a forward-looking rule. A grandfather clause is a special provision in a new rule that makes certain existing situations exempt from it, sometimes for a certain period of time.[20] These clauses often exempt older equipment from the law's provisions. For this reason, old equipment can be used strategically by people who don't want to comply with rules that exempt old equipment from the law's reach. People convicted of felonies are generally barred from possessing guns—but in some states, they can possess *antique* firearms manufactured before 1899, which are more likely to be treated as collectibles and are more difficult to use as weapons. Classic cars are often exempt from emissions testing because they are typically not driven frequently enough to have significant environmental impact. To be sure, many people who own classic cars don't buy them *because* they want to avoid emissions tests—but because the carve-out exists, some people do leverage it as a means of avoiding the testing.

There are two ways in which electronic logging rules have allowed truckers to exploit exceptions for old equipment. One was a grandfather clause for *automatic on-board recording devices*, or AOBRDs. Before the ELD mandate, some fleets used AOBRDs voluntarily to keep track of drivers' hours. The AOBRD was similar to the ELD in purpose and function—except that it recorded fewer types of data, and the data that it did capture were much easier to edit manually (and didn't require edit logs to be easy for law enforcement to access, unlike ELDs).

When the federal government mandated ELDs, it didn't want to force fleets that had just installed AOBRDs to immediately upgrade their equipment—so while drivers using paper logs had to install ELDs in late 2017, trucks that already had AOBRDs were given until the end of 2019 to convert. This put AOBRD-using companies in an interesting position: by being forward-looking enough to install *some* monitoring technology, they kept at bay the requirement to use an even stricter system for another two years, and were able to take advantage of the comparative flexibility of the AOBRD for that time.

The second exception concerns old *trucks* rather than old monitors. The FMCSA initially announced that the ELD requirement would apply only to trucks of model year 2000 and later, based on concerns about whether older trucks would be compatible with ELD technology; many old engines didn't have the right ports to connect to an ELD. Since a 1999 truck would be exempt (and could continue to run on paper logs), truckers resistant to the technology—especially owner-operators, who buy and maintain their own trucks—started showing more interest in buying old trucks, or maintaining old trucks they already owned, to avoid being monitored.[21] Shortly after the rule was announced, market analysts noted that pre-2000 trucks were holding their value better than trucks manufactured just after 2000—presumably due, at least in part, to truckers wanting to avoid monitoring.[22]

The FMCSA introduced a wrinkle when it decided that it would look at the age of the truck's *engine*, rather than its chassis, to determine the exemption—after all, the age of the engine was the impediment to monitoring. This decision brought ELDs into an interesting intersection with a raging environmental debate. "Glider trucks" are new trucks with old engines: glider companies buy newly manufactured truck bodies and retrofit them with rebuilt 1990s-era diesel engines from salvage yards. They do this because the used engines make the truck exempt from emissions requirements—and these trucks emit as much as *fifty-five times* more air pollution as trucks with modern engines.[23] (In fact, the trucks are often nicknamed "super-polluters.") Regulators at the EPA have repeatedly tried to close the glider-truck loophole in emissions rules by limiting the number of glider trucks a manufacturer could make each year; President Trump's EPA Director Scott Pruitt removed the cap on his last day in office before resigning in the wake of ethics scandals.[24] In addition to being super-polluters, glider trucks' old engines also make them exempt from the electronic monitoring rules—making them that much more attractive to independent truckers who bristled at both sets of regulations.

So truckers wanting to evade monitoring rules might get around them by using various types of old equipment. But some truckers chose a different route—buying *new* equipment that was purposefully limited in what it could do. These truckers accepted that ELDs were required, but sought out products that met only baseline levels of compliance, with no extra bells and whistles.

This may not seem, at first blush, like resistance. Minimal compliance is still, technically, compliance, even if just barely. But the marketing of minimally compliant ELDs tells us something about what truckers who bought these systems were trying to say by doing so. Companies that sold these monitors took a very different approach to marketing than other companies.

FIGURE 6.3. The F-ELD's advertising campaign.

While most companies talked about all the capabilities their ELDs had and how much useful data they could generate, these companies acknowledged how unwelcome they were in drivers' lives. ISE Fleet Services, for example, told truckers: "[w]hen you've run hundreds of thousands, or even millions of miles incident-free, validating that you know best when and how you should be operating a CMV [commercial motor vehicle] it's easy to see the ELD Mandate as a disrespectful intrusion. At ISE Fleet Services, we understand that."[25] The ELD system it sells captures "the information necessary for compliance and nothing more."

But perhaps no product captures the resistance of the minimally compliant monitor like the suggestively named "F-ELD." The company's CEO said that it designed the monitor to "push the boundaries of the mandate" through design decisions meant to put drivers' interests first.[26] For example, the F-ELD didn't highlight hours violations automatically like some other models did, in order to make it that much more difficult for law enforcement to detect them. Publicity around the rollout of the F-ELD included free beer at trucker trade shows, a series of commercials talking about "Big Brother" and the overregulation of the industry, and the launch of an energy shot "made for truckers." But nothing summed up the F-ELD like a truckside ad it displayed at the Great American Truck Show. The ad read: "We can't put our finger on it. But the ELD mandate just doesn't feel right" . . . alongside an image of a middle finger, presumably that of a trucker flipping off the government for its overreach.[27]

Resisting around the Technology

Some strategies for beating the box might seem to have little to do with the box itself. That is, the unwelcome presence of the ELD induces truckers (and others) to conduct their affairs differently than they otherwise would—but not in ways that involve directly tampering with the box, manipulating its data, or exploiting its technical limitations. Instead, surveillance drives what we might think of as *second-order* resistance strategies—the reorganization of relationships among firms, truckers, and other players *based on* the technology, but without making any attempt to alter its functionality.

EXITING THE FIRM OR THE INDUSTRY

Perhaps the most direct form of dissent is simply to "vote with one's feet":[28] to quit one's job at a company that is deploying monitoring technologies. In trucking, quitting one's job is not as personally disruptive as it could be in another occupation. In more typical workplaces, people build up auxiliary social commitments around their jobs—office friendships, daily routines, nearby childcare, and the like. The sociologist Howard Becker called entanglements like these "side bets," and theorized that they help to explain why people commit to particular organizations over the long term.[29] But because drivers are mobile and mostly work alone, they often don't have the same contingencies attached to the firms they work for. Moreover, trucking is a highly transferable skill—trucking for one firm is nearly indistinguishable from trucking for another—and the nationwide driver shortage enables drivers to be hired by another company immediately. So, truckers are not particularly brand-loyal employees; as discussed in chapter 2, the rate of turnover in the industry is extremely high, in part driven by drivers bouncing from firm to firm.

This said, departure from a firm can serve as a form of resistance against that firm's electronic monitoring policies. Initially—before ELDs were mandated by the government—a number of drivers told me that they had left firms when those firms installed them, or that they *would* quit if their current carrier did so. Indeed, trucking companies recruiting in truck-stop media sometimes explicitly advertised that they *didn't* require their drivers to use electronic logs. Following the mandate, firm exit may be a less effective strategy—though as we have seen, firms still vary in how intensively they monitor and manage their fleets.

FIGURE 6.4. Advertisements for trucking firms promote lack of electronic monitoring as a recruitment tool. "EOBR" and "Qualcomm" refer to electronic logging devices.

Some drivers I spoke to pre-mandate told me they would quit trucking altogether if the mandate came to pass. As one driver put it in a regulatory comment:

> As a truck driver and otherwise employed in the industry since 1966, during which time I have not killed or injured anyone, I will tell you this: The day the government puts an [ELD] on my truck will be the the [*sic*] last day I ever drive a truck. I know I am not alone in feeling this way. Most drivers I talk to share these feelings. We have to abide by enough rules and regulations that no other motorist deals with without this added insult.

Though not all of these truckers likely made good on their threat of departure, history suggests that drivers *do* respond to undesirable policy changes through industry exit. In 2013, FMCSA tightened rules related to drivers' hours of service by further limiting the hours truckers could drive and mandating additional break periods—changes many drivers considered to be objectionable and unnecessary constraints on their productivity. In the ninety days before the new rules took effect, Werner Enterprises—one of the nation's largest trucking firms—reported that the number of drivers between sixty and sixty-seven years old driving for the firm plummeted by half.[30] After the ELD mandate took effect, one fleet owner said that he had to "talk [drivers] off the ledge" of quitting their jobs, telling them: "You just have to change your mindset and your expectations, and you'll be fine. You can't drive these trucks like you're on paper logs anymore."[31] Other industry players reported that, despite speculation that up to 5 percent of drivers might "hang up their keys," no mass exodus materialized after the mandate.[32]

The departure of older drivers is particularly concerning from a public safety perspective. These drivers, who tend to be the most vehemently opposed to electronic logging, also have more industry experience and are involved in accidents at lower rates than younger drivers. One federal study found that drivers with less than five years of experience were 41 percent more likely to have caused an accident than those with at least five years' experience—and the risk continued to decline as drivers spent more years driving. Twenty-five-year veterans were the least likely drivers to cause accidents.[33] Should these "old hands" leave the industry in droves in response to electronic monitoring, their jobs will presumably be filled by new "baby drivers" (in trucker parlance) who are *more* likely to be involved in accidents.

THE SLOW ROLL

In early 2019, some truckers took a different tack, taking their ELD frustrations to the streets—specifically, to the I-465 beltway that loops around Indianapolis. Truckers used a twenty-five-thousand-member Facebook group called #BlackSmokeMatters (an offensive reference to #BlackLivesMatter protests also occurring across the country at this time) to organize a "slow roll" on February 12: a protest in which truckers would drive the minimum allowable highway speed, forty-five miles per hour, as a way to raise awareness among the general public about the ELD requirement and what they saw as government overreach.

Truckers have used political tactics like these in the past. In the 1970s, they organized convoys and shutdowns to protest rising gas prices, using CB radio to communicate and organize. One shutdown in December 1973 created a twelve-mile traffic jam after hundreds of big rigs stopped in the middle of I-80 through Pennsylvania. The biggest strike, in February 1974, led to major disruptions across the economy—as produce went bad, slaughterhouses ceased operation, and auto plants laid off workers since new cars couldn't be moved. That strike eventually turned violent, prompting the activation of National Guard troops across the country as tensions rose. After eleven days, most truckers went back to work, having met with some success—the federal government allowed independent truckers to add a temporary fuel surcharge to their fees, guaranteed fuel availability at truck stops, and gave national visibility to their woes. The end of the strike made the front page of the *New York Times*.[34]

Perhaps truckers had the 1970s heyday in mind when they planned the Indianapolis slow roll. Thousands of truckers had taken part in those strikes, and Indiana State Police anticipated four hundred to five hundred trucks to snarl traffic that day. But in the end, the protest drew merely seventy-eight trucks—delaying traffic only slightly.[35] A few weeks later, the Facebook group organized another slow roll in Chicago, with similarly meager turnout. Organizers had thought the slow rolls could set the stage for a nationwide shutdown. But they called off those plans, acknowledging that the numbers "were nowhere near what we needed" to make an impact.[36] "We lost the fight," one of the group's founders said, noting that he was "disappointed and heartbroken" that the slow roll wasn't more effective as a way to resist electronic monitoring.

COLLABORATIVE OMISSION

A particularly interesting resistance strategy involves behind-the-scenes cooperation between truckers and a surprising third party: truck stops. Truck stops' stake in the use of electronic monitoring in trucking is not immediately obvious. But they *do* have an interest in competing for truckers' business, and they may exploit truckers' desire to circumvent the timekeeping regulations as a means of doing so.

In addition to the primary timekeeping rules, truckers are subject to a regulation called the "supporting documents" rule.[37] When a law enforcement officer inspects a driver's time logs—in paper or electronic form—they ask the driver for documents that verify his recent whereabouts. Truckers are required to keep these documents to help the inspector assess whether the driver's logs are accurate with respect to when he was at a given location. The kinds of documents truckers can use to fill these requirements could include bills of lading, toll slips, inspection records, and—importantly—receipts for fuel or other services at truck stops. Because truckers stop at truck stops frequently to refuel, buy food, take showers, or for other reasons, drivers commonly keep such receipts in the cab.

Knowing that many truckers have an interest in "flexibly" logging their time, some truck stops began providing receipts that list a location and date—*but not an exact time*—at which a transaction occurred. Doing so prevented an inspector from using the document as a "tattletale" if the trucker's paper or electronic log would not corroborate the time of the transaction, and gave him a bit more leeway to "fudge" a paper log (or, to a far lesser extent, an electronic log). This provided an incentive for truckers to patronize those truck stops that provide such documentation. (In 2017, FMCSA tightened the requirements for what truckers could count as a supporting document, making this practice somewhat more difficult than it had been previously—but in some circumstances, non-time-stamped receipts can still be used to support a driver's claims.)

We might think of this form of resistance as *collaborative omission*. The strategy depends on the actions of multiple parties, some of which are external to the primary enforcement relationship. And unlike some of the other forms of resistance discussed, such as material tampering, this tactic appears more passive; the trucker is merely doing what is required of him by providing supporting documentation, while maintaining the ability to disclaim the intent to mislead. In fact, surely, many truckers provide non-time-stamped receipts *without* any resistant intent.

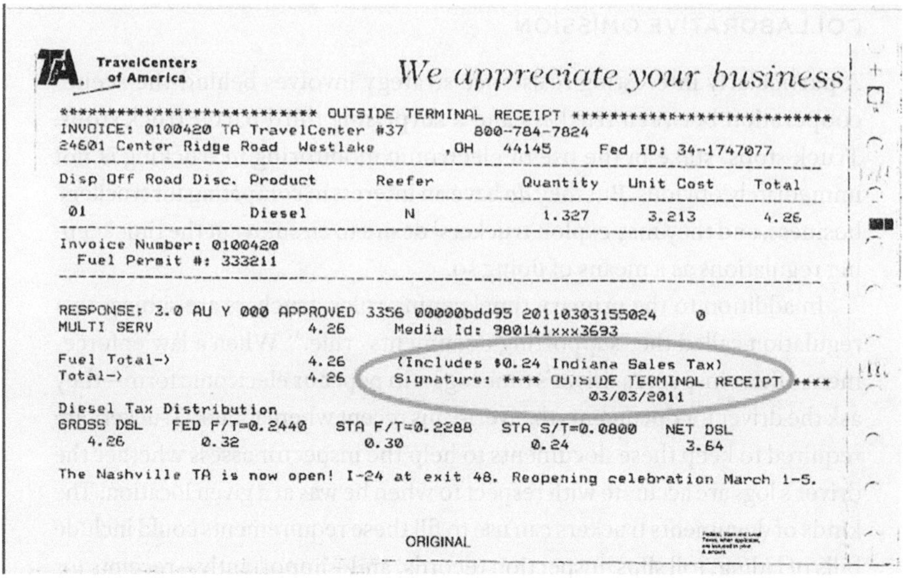

TravelCenters of America

We appreciate your business

************************** OUTSIDE TERMINAL RECEIPT **************************
INVOICE: 0100420 TA TravelCenter #37 800-784-7824
24601 Center Ridge Road Westlake ,OH 44145 Fed ID: 34-1747077

Disp	Off Road Disc.	Product	Reefer	Quantity	Unit Cost	Total
01		Diesel	N	1.327	3.213	4.26

Invoice Number: 0100420
Fuel Permit #: 333211

RESPONSE: 3.0 AU Y 000 APPROVED 3356 00000bdd95 20110303155024 0
MULTI SERV 4.26 Media Id: 980141xxx3693

Fuel Total-> 4.26 (Includes $0.24 Indiana Sales Tax)
Total-> 4.26 Signature: **** OUTSIDE TERMINAL RECEIPT ****
03/03/2011

Diesel Tax Distribution

GROSS DSL	FED F/T=0.2440	STA F/T=0.2288	STA S/T=0.0800	NET DSL
4.26	0.32	0.30	0.24	3.64

The Nashville TA is now open! I-24 at exit 48. Reopening celebration March 1-5.

ORIGINAL

FIGURE 6.5. Truck stop receipt displaying date, but not time, of transaction. Source: https://www.apexcapitalcorp.com/fuel/indiana-sales-tax/.

DECOY COMPLIANCE

Finally, recall our discussion in chapter 5 about truckers' pre-mandate use of truckside stickers, playing upon human law enforcement officers' perceived reluctance to inspect ELDs. As described, truckers who engaged in this practice relied on the hope that inspectors, who have discretion as to which truck to pull into a bay, would choose not to select them for inspection. This *simulation* of compliance had led to a "black market" for decals. Decoy compliance can be understood as another form of resistance to electronic monitoring. By signaling that one is likely compliant, a trucker is afforded "cover" that enables him to break the rules.

Resistance as Networked Negotiation

It is almost too obvious to observe that these resistance practices demonstrate that electronic monitoring doesn't take noncompliance "off the table" (as at least one trucking company vice president opined). It may be harder to break the rules with an electronic monitor than it was with pencil and paper—but given the right knowledge, resources, and motivations, it is still possible.

The dozen resistance practices I've described look very different from one another. They vary from blunt smashing to digital tinkering to social signaling; from overt, communicative action to covert maneuver; from solo endeavor to coordinated effort. Resistance can be passive (this is just the receipt the gas station gave me!) to indisputably active. Data can be falsified, omitted, obfuscated—or its collection thwarted in the first place. Resistance can be a way to express "this is who I am!" or a way to say nothing at all, when flying under the radar is the best way to accomplish one's goals. Some forms of resistance are primarily directed against law enforcement, while others respond to employer-originated surveillance. The tactics vary by target (occupational performance monitoring versus legal compliance oversight versus both) and by principal aim, from preservation of occupational identity to financial gain to compliance with company rules.

The variation is the point. Resistance against surveillance in trucking is multidirectional and multivalent: it pushes against multiple forces in many different ways, marshalling whatever resources are available—a receipt, a hammer, a decal, a spare password, a loophole—to accomplish a wide variety of aims for a wide variety of actors.[38] And each involves different domains of knowledge, from technical (this is the fuse to disconnect) to organizational (this is the miles-per-hour threshold set by the trucking company, below which the ELD won't register me as "driving") to social (these are the skills and anxieties of inspectors).

Most strikingly, the trucking case complicates the conceptualization of resistance as being a counterstrategy against authority writ large. It is too simple to assume, as much thinking about resistance does, that resistance is by necessity a "bottom-up" phenomenon—that it is user-motivated, or even, for that matter, user-empowering. Even though the imagery of the lone-wolf trucker vigilante is often invoked, each of the strategies detailed in this chapter implicates *multiple* actors acting in relation to one another. Trucking firms, even as they are surveillors themselves, can also act as resistors against government monitoring, by tinkering with back-office data or instructing their employees about how to game technical thresholds. Even seemingly disinterested third parties get into the mix, as truck stops offer receipts that enable truckers to log their time more flexibly. Instead of being an oppositional, bottom-up, "weapons of the weak" response to top-down power, resistance strategies become an avenue through which stakeholders attempt to arbitrate power relations among themselves, both upwards and downwards. In this way, resistance brings actors into temporary regimes of alignment[39] and collaboration in order to enact maneuvers that *may or may not* serve the ultimate ends (social, economic, cultural) of the party who appears to be the chief "resistor" (here, typically, the truck driver)—but which do serve as tactics for the negotiation of relational power.

For example, in one (albeit nonrepresentative) industry survey, about half of truckers said they felt *more* pressure from carriers, brokers, and shippers to cheat on hours regulations since the ELD mandate.[40] This may be because before, these parties treated the rules more as "guidelines" with built-in flexibility, and assumed that truckers would do what they needed to in order to get the job done—without directly pressuring truckers to do so. It may be that because the new rules are less flexible, some companies feel they need to push drivers more explicitly to "beat the box"—after all, their business still depends on it.[41]

Making Sense of Self-Exploitation

Understanding resistance in this way helps us to reconcile a paradox we alluded to in chapter 2: the idea that, through some acts of resisting (driving with "ghost logs," decoy compliance, mis-logging nondriving hours, etc.), truckers in fact *reinforce* economic structures that serve to exploit their labor and jeopardize their health and safety. When truckers' resistance stands in opposition to the hours-of-service limitations, resistant truckers are simply enabling themselves to work longer, more dangerous hours (during which they put themselves and the motoring public at risk), in conditions comparable to sweatshop labor.[42] So even seemingly trucker-initiated resistance can be understood, in a sense, to be self-exploitative: truckers' resistance often serves merely to reproduce and entrench the economic and class structures that underlie industrial capitalism. By doing so, truckers "take a hand in their own damnation."[43] And as described in chapter 3, truckers' resistance, though often painted with a cultural veneer of cowboyism and machismo, is a function of bare economic necessity (recall the trucker who reminded me that for many truckers "the lights wouldn't be on at home if they weren't breaking the law").

But this situation becomes more comprehensible when we take a broader view of truckers' resistance—who resists, who benefits, and why—and recognize it as a means of negotiating networked relationships. The trucker who drives with a ghost log is not merely acting in opposition to the law. He is also actively negotiating his relationship with his employer, who may have provided the resources for this subterfuge, and whose exerts significant economic influence (and possibly even direct coercion!) over him. In addition, we must consider *who benefits* when resistance practices are used. In some of the cases I describe—for instance, the dispatcher instructing the driver to "roll" to a terminal, or the mis-logging of inspections and detention time—the trucking *firm* is the beneficiary of the resistance practice; the trucker himself may not even be willing to participate in it, but may be coerced into doing so.

Truckers whose resistance practices are primarily intended to thwart surveillance from the firm are also engaged in networked negotiation. When a trucker disrupts his GPS receiver using an aftermarket jammer, or disconnects a fuse after a fellow trucker has instructed him in how to do so, he is often aiming to get extra driving time, which means he can earn more money from the company. But in many cases, it also cements his identity in relation to his community. It can serve a collectivity-defining interest (perhaps especially if the practice is overtly communicative, as in the coordinated hammer-smashing case), or, even if executed individually, connects him to the longstanding cultural values of his occupation: independence, the capability to tinker with machinery, defiance at being told by an authority figure how to conduct one's work, and affiliation with a pool of knowledge resources that can be mobilized and communicated to other truckers. This view leads us to reconsider what the aims of resistance are—or what resistance is *for*, even in the absence of institutional change or collective action.

As Patricia Ewick and Susan Silbey suggest,[44] everyday acts of resistance become sociologically significant when they construct *meanings and narratives* for their participants which lay bare their relation to structures of power. Such dynamics are evident in the trucking case—as resistance practices, even if economically self-exploitative, provide resources for truckers to make sense of and negotiate the occupational communities and institutional networks in which they are located.

Hollow Victories?

It's impossible to characterize truckers' resistance as simply "good" or "bad." When truckers resist to get extra drive time and make more money, we might view this as an important way to preserve their autonomy and economic well-being—but also, potentially, as an exercise in self-exploitation. When truckers resist to assert their identities, but without improving their economic standing or thwarting systemic surveillance in a significant way, we might see this as a meaningful mode of self-definition—or as a hollow, palliative exercise that fails to challenge the system. As resistance scholars

Ashley Rubin[45] and Torin Monahan[46] have noted, some forms of resistance can, perversely, have the consequence of further entrenching surveillant systems (for example, by giving surveillors narrative ammunition to support harsher punishments for noncompliance).

Scholars of legal and political resistance have conceptualized small acts like these in different ways. Ewick and Silbey document tactics they call "anti-discipline"—"dodges, ruses, and feints [that] rarely leave a structural imprint"[47] on systemic orders. But they caution us not to dismiss small resistant acts as trivial. These acts may, as James Scott theorized, open the door for bigger, more consequential challenges to power later on. But even if they don't, small resistant acts are important in their own right because they have meaning to the person carrying them out. As Ewick and Silbey put it: "[T]he moments of resistance are often the most memorable parts of the journey. To ignore [these] tactics because they are momentary and private is to reinscribe the relations of power they oppose. . . . To dismiss these momentary feints and ruses is to deny the dimensions of [one's] identity forged in the cracks of the law."[48]

7

RoboTruckers

THE DOUBLE THREAT OF
AI FOR LOW-WAGE WORK

So far, this book has focused on the trucking workplace as a site of intensifying surveillance, within which workers are much more closely monitored and managed than they have ever been before. We have seen how electronic monitoring challenges truckers' professional culture and autonomy, as well as their ability to make the money they need to get by, and how truckers have responded to this technology. Over the last fifteen years, as monitoring technologies have become ubiquitous, the specter of putting the manager (and the government) in the cab is gradually becoming an unwelcome reality for many truckers.

But recently, another narrative has emerged alongside this one. Economists and policymakers are becoming increasingly concerned about the effects of automation and artificial intelligence on employment—including whether some kinds of jobs will cease to exist at all. Trucking is often thought to be one of the first industries at substantial risk from the prospect of increasing automation, as self-driving trucks seem to be on the horizon of technological possibility. This leads to a newfound concern about the relationship between trucking and technology: what will become of truckers if they are no longer needed behind the wheel? This chapter explores that threat.

Automating Work and Workers

New technologies can profoundly affect how industries are organized and how work gets done within them. As AI technologies become cheaper and more sophisticated, policymakers have begun to express great concern about what they will mean for employment—including whether some forms of work will exist at all, and what will happen to the workers who do that work. The prospect of millions of workers finding themselves suddenly without employment obviously brings with it the potential for tremendous social and economic disruption.

The jobs widely believed to be most threatened by the new wave of AI-led automation are those commonly held by less educated, poorer workers—what are often called "low-skill" jobs. "Low-skill" is an ambiguous (and somewhat misleading) term, as many jobs categorized as "low-skill" inarguably require significant expertise and ability. "Skill" is often used as a sort of shorthand for the amount of education or specialized training that a job requires. When economists refer to "low-skill" jobs, they often refer to blue-collar jobs that require no more than a high school or vocational education. "High-skill" jobs, by contrast, often require college or a more advanced degree.[1]

Earlier waves of technological innovation—like computer-controlled machinery on the factory floor—have been *skill-biased*: that is, automated technologies generally replaced jobs held by low-skill workers. On the other hand, workers with more education were both less likely to be threatened by technological replacement, and in some cases were in even greater demand, as new technologies created new types of jobs for them. In this way, innovation has had a disproportionate negative impact on the labor of those at the bottom of the socioeconomic ladder.

What is it about a job that makes it more likely to be automated? Historically, one of the primary axes of distinction between automatable low-skill jobs and safer high-skill jobs was whether work consisted mostly of *manual* or *cognitive* tasks. Machines were comparatively better at manual (physical) tasks, while cognitive tasks—"knowledge work" involving processing, analyzing, or acting upon information, which is more likely to be held by highly educated workers—were largely beyond machine capabilities, and remained within the ambit of the human worker. But over time, computers became much more capable of analytic information-based work, making the distinction between manual and cognitive work tasks less salient. Instead, the more crucial distinction became whether a task could be easily boiled down into a standard, repeatable method. In 2003, the economists David Autor, Frank Levy, and Richard Murnane published an influential model that categorized work tasks not only as manual or cognitive, but also as *routine* or *nonroutine*.[2] A routine task is one that requires "methodical repetition of an unwavering procedure"[3]—the task involves doing the same thing over and over, like fixing the same part into place on an assembly line or assisting a customer with an everyday request. Nonroutine tasks, on the other hand, have less clearly understood rules and procedures—and therefore, the economists predicted, would be less susceptible to being programmed away. Autor and his coauthors argued that computers might *complement* workers in nonroutine tasks but would be less likely to substitute for them. Nonroutine tasks—like selling a product or writing a legal brief—require skills like perception, problem-solving, and intuition that were (at the time) well beyond the purview of what a computer could reliably accomplish. Truck driving, in fact, was one of the jobs that Autor and his coauthors then believed would be *safe* from automation, because the nature of truck-driving work involves dealing with constant nonroutine scenarios.

But the latest wave of AI-driven technological innovation is changing the game once again. As computers become more sophisticated and responsive to their environments, they can adapt to dynamic situations more adeptly— negotiating traffic, responding to conversational cues, developing novel solutions to problems. And this suggests the possibility of a new threat to workers whose tasks are primarily *non*routine. In light of robotic capabilities, computer vision, and machine learning, it's less important than it once was that a task be clearly definable and repeatable. Today's algorithmic systems are fed by tremendous amounts of data, can recognize and interpret speech, and can predict some types of outcomes far better than a human can. They can complete myriad nonroutine tasks, both manual and cognitive: they may offer personalized financial advice, diagnose medical conditions, navigate new environments, and assist customers with complex needs. And though these systems certainly make errors and have nontrivial limitations, they are capable of "learning" constantly—integrating new information from the environment, from new data sources, and from past mistakes.

No one knows for certain just how this wave of technologies will impact workers, but some economists project that the situation will be dire, particularly for low-wage workers. In an influential 2013 paper called "The Future of Employment: How Susceptible Are Jobs to Computerisation?"[4] economist Carl Frey and machine learning researcher Michael Osborne predicted that about 47 percent of total US employment was at risk of being automated within the next twenty years—and that low-skill, low-wage jobs, in sectors like office administration and telemarketing, were likely to be hit hardest. In 2016, the White House's Council of Economic Advisers projected that jobs with wages under twenty dollars per hour were 83 percent likely to be automated, using the Frey & Osborne data, while those with wages over forty dollars per hour were only 4 percent likely to be automated.[5]

Robot, Take the Wheel?

Based on these economic projections, trucking seems to be a prime target for automation. A 2016 White House Report on Artificial Intelligence, Automation, and the Economy forecast that 80 to 100 percent of heavy and tractor-trailer truck driving jobs (between 1.3 and 1.7 million jobs) would be "threatened or substantially altered" by autonomous vehicle (AV) technologies.[6] The Frey and Osborne report concurred, estimating that truck driving was about 79 percent likely to become automated in the next two decades.[7] Other economic researchers make similar predictions, including Goldman Sachs (which estimated in 2017 that trucking stood to lose about three hundred thousand jobs per year)[8] and the International Transport Forum (which projected 50 to 70 percent reduced demand for drivers in the United States and Europe by 2030).[9] The threat is commonly described in calamitous terms: as a "revolution,"[10] or as the "death of the American trucker."[11] As Doug Bloch, political director for the Teamsters, put it: "everywhere I go I hear people talking about a robot apocalypse, about four million transportation workers losing their jobs in the next five to twenty years."[12]

As we've seen, trucking may seem to be an industry ripe for disruption. The work itself is difficult, unsafe, and often deadly. Major inefficiencies in the industry—including the astronomical rate of driver turnover—could be ameliorated if human drivers were just not needed at all. Industry problems and the difficult nature of trucking work create strong "push factors" in favor of an automated solution. And of course, the same concerns about driver fatigue and overwork that motivated electronic monitoring in the industry (discussed in chapter 3) may also serve as justification for making trucks drive themselves: machines can work long hours; they don't get sleepy, and they don't need to take breaks. The tension between the legal restrictions on drivers' work hours and the speed of business demanded by the industry might be significantly eased if machines could do more of the work. Negative stereotypes about truckers play a role in justifying the move to autonomous trucks, too. One trucking tech executive notes that a customer observed that "a robot is unlikely to get into a fight with people at a warehouse, or decide to stop in Las Vegas and spend a few days in a strip club."[13]

There are significant cost considerations, too. Some industry forecasters claim the autonomous revolution will be the biggest economic disruption to trucking since deregulation in 1980, or even since the launch of the American highway system in the 1950s.[14] In a best-case scenario, fully automated trucks could save companies money in lots of ways. Automated trucks would use fuel more efficiently (largely because of the possibilities for *platooning*, in which trucks drive very close together to reduce wind drag); insurance rates might decline; and asset utilization—the amount of time the trucks a company owns are actually on the road, driving, rather than sitting unused in a parking lot—could increase dramatically. And of course, most significantly, robots don't need to get paid. If autonomous trucks reduce demand for human labor—a big "if," as we'll see—companies could save millions of dollars in wages currently paid to human drivers. Some estimate that these cost savings to companies could add up to more than a dollar per mile driven—about half of which would come from reduced labor costs.[15]

So, trucking might be perfectly situated for automation. There aren't enough human truckers to meet demand, and the truckers who are on the road are faced with unsafe and unhealthy conditions. And trucking companies might see significant savings if they can operate their equipment more efficiently, and without paying humans to do it. As one autonomous trucking executive put it: "Trucking needs self-driving and self-driving needs trucking."[16]

Driven by these factors, driverless trucks have become a site of tremendous technical innovation and investment. Machine learning, sensing, and object recognition have improved an enormous amount over the last decade, leading to major investment in self-driving vehicle research and development. And these systems are currently even better suited for trucking than for passenger cars: autonomous vehicles typically do their best driving on highways—where the great majority of trucking miles are driven—rather than on city streets, where speeds are more variable, pedestrians more frequent, and obstacles less predictable.[17]

Accordingly, investors are making big bets on autonomous vehicles. A 2017 Brookings report stated that autonomous vehicles were "the leading edge" for investment in AI systems, and projected continued market growth into the future.[18] McKinsey reported $120 billion of carmaker investment in autonomous vehicle technologies, from sensors and semiconductors to AV mapping software, from 2017 to 2019.[19] And one market research study predicted that the global AV market would increase tenfold between 2019 and 2026.[20]

Alongside these investments, companies have begun publicly touting the accomplishments of their autonomous trucks. Perhaps the most poetic of these was the well-publicized 120-mile test delivery, in Colorado, of fifty thousand cans of Budweiser via autonomous truck. The test was successful—though the truck had a safety driver, and was accompanied by a convoy of four state patrol cars, three company vehicles, and two tow trucks—a level of support clearly unsustainable for everyday beer deliveries.[21] Autonomous trucking startup Embark similarly publicized its successful transport of refrigerators from El Paso to Palm Springs.[22] In Europe, a convoy of self-driving trucks platooned across the continent, providing a public demonstration of the technology's potential (organized by the Dutch government).[23] Though these test runs and pilot projects are still accompanied by human backup drivers, the near-term goal is to remove the trucker from the cab altogether, at least for the majority of the time the truck is operating.

Systems like these conceive of the trucker as a *displaced* body. He is displaced both physically and economically: removed both from the cab of the truck and from his means of making a living. As former Google engineer Anthony Levandowski envisioned it in 2016, drivers would still be required in the short term, but "long down the road, none of the new trucks will have a cab on them. It just doesn't make sense."[24] In this narrative, the autonomous technology has replaced the human trucker's imperfect, disobedient, tired, inefficient body, rendering him redundant, irrelevant, and jobless. And the result of this displacement, it is feared, could be sector-wide unemployment and the rapid dissolution of up to two million workers' livelihoods.

A Slope, Not a Cliff

But the reality of the situation is far more complicated. To be sure, technology-driven unemployment is a real threat, but robotic trucks are very unlikely to decimate the trucking profession in one sudden phase transition. The path to fully autonomous trucking is likely to be a gradual slope, not a steep cliff—a trajectory shaped not only by technical roadblocks (more on this in a moment), but by human, social, legal, and cultural factors, which economic forecasts about AI and job loss commonly bracket from consideration.[25] Sociologist Steve Viscelli warns against committing a "Jetsons fallacy" when it comes to autonomous trucks—that is, forecasting the future based on technical capabilities in isolation, without considering how those capabilities will intersect with social factors. (Recall that the 1960s cartoon "The Jetsons" envisioned highly futuristic technological developments, like robots and flying cars—but superimposed them on the culture and social structure of mid-century America.)[26]

Truck drivers' daily work consists of much more than driving trucks. Truckers monitor their freight, keeping food at the right temperature in refrigerated trucks and loads firmly secured to flatbeds. They conduct required safety inspections twice a day. They are responsible for safeguarding valuable goods. They maintain the truck and make repairs to it—some of which are routine, and some less so. When truckers arrive at a terminal or delivery point, they don't just drop things off and leave: some load and unload their freight; they talk to customers; they deal with paperwork; they may spend hours making "yard moves" (waiting for an available delivery bay and moving to it, much as planes do at busy airports). Could some of these tasks be eliminated by intelligent systems? Surely some can and will—but these components of the job are much harder to automate, and will come much later, than highway driving.[27]

And this is true of all work. Though we may boil jobs down to their component tasks for purposes of analyzing and comparing them, anyone who's held a job knows that work depends on deep-seated human knowledge that cannot always be boiled down to rulesets and protocols. The philosopher Michael Polanyi called this the *tacit dimension* of human knowledge[28]— there are things humans know and do that evade easy categorization, and can barely be articulated, let alone automated. As Polanyi put it: "The skill of a driver cannot be replaced by a thorough schooling in the theory of the motorcar; the knowledge I have of my own body differs altogether from the knowledge of its physiology; and the rules of rhyming and prosody do not tell me what a poem told me, without any knowledge of its rules."[29] On Polanyi's reading, human activity is more art than science. Even if we can offload some tasks to machines, machines can never really replace people, precisely *because* they are machines.

Machines are increasingly better at learning the rules—but have a hard time knowing when rules should *not* be followed, or how to balance among competing demands in safety-critical situations. Just as ELDs make it more difficult for truckers to bend the timekeeping rules, autonomous vehicles may struggle with knowing when to bend the rules of the road, and which ones. Indeed, some accidents have been caused by AVs hewing *too closely* to the rules (for instance, driving the posted speed limit). But driving requires fluidity, attention to local custom, and communication with surrounding drivers and pedestrians—tacit knowledge that may be impossible for robots to fully grasp, but which is essential for the social coordination of driving amidst human beings.[30] As Missy Cummings, director of Duke's Humans and Autonomy Lab, puts it, autonomous cars "don't understand social graces."[31]

Automation tends to threaten particular work *tasks*, but rarely replaces entire jobs.[32] One oft-repeated example is the bank teller. Despite widespread fears that ATMs would replace the job of the human teller in the 1990s and 2000s, teller jobs not only increased, but increased *faster* than the rate of the general labor force. Tellers' jobs changed—they now do more marketing and customer relations work than plain cash-handling—but they didn't disappear.[33] There are, of course, counterexamples: "pin boys" no longer reset the pins at bowling alleys, and there are far fewer elevator operators than there once were. But jobs that consist of many and complex tasks are unlikely to be automated away, as there will always be parts of the work that machines can't do.

Social factors may slow adoption, too. The public fears autonomous vehicles. In a 2017 Harris poll,[34] 69 percent of Americans thought riding in an autonomous car sounded very or somewhat risky; a Gartner poll the same year found that 55 percent of people would not ride in a fully autonomous car, and 29 percent would not ride in a car that was partially autonomous (allowing the driver to take over if necessary).[35] Drivers are nervous about issues like hardware and software malfunctions, data security and hacking, and loss of control over the vehicle.[36] And though there have been relatively few crashes and fatalities involving autonomous vehicles to date (particularly in comparison to rates among nonautonomous vehicles), those that have occurred have attracted significant media attention and public response. When an Uber AV struck and killed a pedestrian in Arizona in March 2018, Jason Levine, executive director of the Center for Auto Safety, said that the crash "set consumer confidence in the technology back years if not decades."[37] This crisis in confidence is amplified by stakeholders with vested interests—like the international president of the Transport Workers Union, who called the Arizona accident "a stark reminder that our elected officials should think long and hard before they put dangerous humanless RoboBuses on the streets[.]"[38] Later that same week, a Tesla in "Autopilot" mode struck a highway divider in California, killing its driver and sending Tesla's shares tumbling. Tesla CEO Elon Musk responded by blaming human drivers for being overly "complacent" about Autopilot mode and not paying sufficient attention while it was engaged.[39]

Given public apprehension about autonomous cars, autonomous trucks may face even more hesitation. The prospect of an opaquely controlled, hundred-thousand-pound truck barreling unmanned down the highway is, for many, the stuff of nightmares or horror films. (Not for nothing, Stephen King's 1986 film *Maximum Overdrive* features an autonomous truck with a life of its own and a taste for blood.) As the head of the American Transportation Research Institute put it, "driving next to any car or truck, and not seeing a human in it, is certainly going to freak out Grandma."[40] Public opinion will probably adjust somewhat as autonomous vehicles become safer and more commonplace. But this process is likely to take time, and to temper the pace at which autonomous trucks might take over truckers' jobs.

Finally, legal hurdles are also likely to slow adoption of fully autonomous vehicles. New types of vehicles will require a host of new rules, from safety standards and traffic laws to insurance regulations and liability regimes. Standards for vehicle-to-vehicle communication will need to be developed, so that autonomous cars can communicate with one another and with the infrastructure that surrounds them.[41] Steps will need to be taken to ensure that vehicle technologies are secure from hackers and other threats. Though rules of this sort have changed before and can change again, doing so takes time, invites competing interests, and is likely to delay the technical progress of autonomous vehicles for at least some time.[42]

A comprehensive 2018 review[43] of the body of trucking regulations assessed how compatible these rules might be with autonomous trucks. The review found that a number of rules and assumptions would need to be reassessed in light of autonomous trucks. For example, current rules require that a driver examine the cargo on his truck every three hours or 150 miles to ensure it is well-secured;[44] it is not immediately obvious how tasks like these are to be carried out without a human driver. Similarly, should current medical restrictions, licensing requirements, and hours-of-service rules apply to humans working in tandem with autonomous trucks, or remotely? It is not immediately clear.

Another issue is coordination among the hodgepodge of state regulations that control new vehicle technologies on public roads. As legal scholar Bryant Walker Smith puts it, "[o]n the question of vehicles that don't quite look like a car (or quack like a truck), state and federal law are terrifically muddled."[45] States' conditions for AV testing are currently quite variable. Some states have lenient requirements (in part to lure AV companies to locate within them), while other states have stricter standards that limit AV testing. This isn't a workable long-term situation for autonomous trucks: if an autonomous truck can cruise through Illinois and Ohio but not Indiana, the whole route is effectively stifled.

Eventually, coordination will be necessary to ensure that autonomous trucks can travel easily across state boundaries. This coordination might come in the form of overriding federal laws that preempt the states' patchwork rules. It wouldn't be the first time the federal government has acted to avoid such problems: the Supreme Court has previously invalidated states' patchwork laws regulating truck safety because of the difficulty they posed for interstate commerce. For example, states have previously passed laws mandating that trucks traveling on their roads have curved mudflaps rather than straight ones, or have limited the length of trailers trucks may carry—but state-specific rules like these impede the ability for trucks to legally travel across the country.[46] In each case the Supreme Court has struck down these state laws because of the burden they placed on interstate commerce. Should autonomous trucks become more common, the federal government might adopt a similar approach to ensure that they can travel across state boundaries unimpeded.

There's some indication that the federal government is already aiming to create consistent rules. The National Highway Traffic Safety Administration (NHTSA) has so far taken a light touch in regulating autonomous vehicles—perhaps too light, according to some safety and consumer groups—in the name of fostering innovation. But there are some indicators that NHTSA is beginning to take a more active role in their development, stepping up crash investigations and requiring more reporting from autonomous vehicle companies about their testing.

But lawmaking takes time, and any new rules are likely to face some opposition—not only from workers, but from highway safety interest groups, anxious insurers, and a substantial proportion of the public.[47] Though new regulatory regimes will be developed, they are likely to be introduced and refined over time—another reason that autonomous trucks will arrive on a slope, not a cliff.

All of these factors—the complexity of the tasks that comprise trucking work; social and cultural hesitation around autonomous vehicles, and autonomous trucks especially; and legal hurdles to implementation—are at odds with the projection of a "cliff" of trucker unemployment brought about by AI technology. Instead of thinking about a sudden wave of unemployment, then, we should think about how AI will *change what work looks like* over the long haul. There will still be human truckers for a long time to come—but this doesn't mean that what it means to be a human trucker won't change substantially. Rather than whole-cloth replacement of human truckers, autonomous technologies might require *integration* between human and machines over a long period of time, as truckers are required to coordinate their work—and themselves—with the technology.

There are several possible forms this integration might take. Here, I'll describe three prospects that have emerged in the trucking context: the *handoff, network coordination,* and *human/machine hybridization*. Each represents a way to think about what work might look like in the near future as autonomous technology becomes more prevalent.

The Handoff: Passing the Baton

One vision of the future imagines machines and humans as coworkers. In this model, people and machines "pass the baton" back and forth to one another, like runners in a relay: the worker completes the tasks to which she is best suited, and the machine does the same. For example, a robot might take responsibility for mundane or routine tasks, while the human handles things in exceptional circumstances, or steps in to take over when the robot's capacities are exceeded. In other words, the human and the robot "hand off" the task to one another synergistically.[48]

Human/robot teams hold some promise both because they try to seize on the relative advantages of each—and because the model presumes that humans get to keep their jobs. In fact, some believe that human jobs might become more interesting and fulfilling under such a model, if robots can take on more of the "grunt work" that humans currently are tasked with completing. John Maynard Keynes predicted in 1930 that technological innovation would lead humanity into an era where the most pressing problem would be how to how to entertain ourselves in our abundant free time, newly freed from the demands of economic productivity: "how to occupy the leisure, which science and compound interest will have won."[49] More recently, economist Erik Brynjolfsson raises the possibility of a "digital Athens" in which humans have more time for leisure and creative work, having outsourced the dullest and dirtiest tasks to machines;[50] even former US Labor Secretary Robert Reich posited that technological innovation and universal basic income might "create a future in which robots do most of the work" and people are freed "to pursue whatever arts or avocations provide them with meaning."[51]

In journalism, for example, AI is used to automatically generate millions of routinized articles about corporate earnings reports, weather forecasts, and sports scores.[52] So-called "robot journalists" pull statistics from predictable sources and use natural language processing techniques to turn them into bare-bones, formulaic articles, aimed at readers who just want the facts and want them quickly (for instance, quarterly earnings reports that contain important information for stock traders to act on). By using a service called Automated Insights to write routine copy, the Associated Press increased the number of its quarterly earnings articles from 300 to 4,400—while freeing up its journalists to "use [their] brains and time in more enterprising ways during earnings season."[53] And rather than being seen as a threat, some journalists seem to welcome their robotic coworkers. When the Associated Press began using Automated Insights, TechCrunch business reporter Alex Wilhelm wrote: "if I could offload the most quotidian [reporting tasks], say, crafting a standard paragraph that compares results with forecasts, and save myself a few seconds, I'd be all for it. No one reads TechCrunch, or me I suppose, because our prose when comparing fiscal third quarter diluted earnings per share on a non-GAAP basis to market expectations is especially riveting. But I think that people do seek out reporting and analysis that helps make those numbers mean something."[54] (Of course, not everyone shares Wilhelm's view; while automated journalism doesn't lead to substantial job loss for human journalists, scholars raise concerns about the credibility of AI-written stories, robots' capacity to follow journalistic ethics, and algorithmic incentivization of certain forms of journalistic work over others.[55])

The human/robot team is not an especially farfetched idea for trucking work. In fact, most of us encounter a version of this model every time we sit behind a steering wheel. Modern cars commonly offer some form of technological assistance to human drivers (sometimes called "advanced driver-assistance systems"). Adaptive cruise control is an example: when a human driver activates it, the car automatically adjusts its own speed to maintain a given driving distance from the cars in front of it. Lane assistance operates similarly—some passenger vehicles can self-steer to avoid crossing lane dividers. Both adaptive cruise control and lane assistance are designed as safety features, under the well-founded assumption that human drivers with limited attention spans and perceptive capabilities will fail to brake in time when cars ahead of them slow down, or will let the car drift into an adjacent lane. Parking assistance follows the same logic: some passenger cars can steer themselves into parallel parking spaces, a skill which many humans find notably difficult. All of these are examples of human/machine "teamwork" in everyday driving. The human remains in charge—but passes the baton to the machine for work the machine is better equipped to handle.

These types of automation might seem very distant from the fully autonomous vehicles we've discussed—and they are, technologically. But conceptually, fully autonomous cars that drive themselves completely lie on a spectrum alongside assistive technologies like these—they all ascribe *some* degree of autonomy to the machine. To clarify what *autonomous* means in a given situation, technologists and regulators use a classification system developed by the Society of Automotive Engineers (SAE), an organization that develops professional standards for the automotive and aerospace industries. The SAE defines six "levels" of autonomy for vehicles. At Level 0, No Automation, the human is fully in control of driving the vehicle, though the car may provide notifications and warnings to help them do so. Level 1, Driver Assistance, means that the car can execute specific driving tasks based on information that the car senses about the environment—like the lane control, adaptive cruise control, and parking assistance examples we've discussed—and that the human does everything else.

As the SAE's automation levels increase, the car gradually becomes responsible for more driving tasks. At Level 2, Partial Automation, the car can "self-drive" under certain conditions—Tesla's Autopilot mode, in which the car accelerates, brakes, and steers independently, is the best-known commercial example[56]—but the human is expected to monitor the environment and to take over if necessary. At Level 3, Conditional Automation, an important change occurs: the *car*, not the human, monitors the external environment and prompts the human to take control if necessary. As the SAE envisions Levels 4 and 5—High Automation and Full Automation, respectively—the human has even less role to play. At Level 4, the car is capable of driving itself under certain defined conditions even if the human is prompted to intervene and fails to do so, and at Level 5, the human need play no role whatsoever in driving.

The most advanced semi-automated technologies on the market today are just beginning to approach Level 3, in which the car does most of the required monitoring of the environment. This means that the machine does most of the work of looking out for and recognizing obstacles, keeping track of the locations and speeds of other cars, and the like—*but* the human is required to "respond appropriately to a request to intervene" (that is, to act in a way that resolves the risk of an accident). In other words, the machine has the baton most of the time, but the human has to be prepared to grab it immediately when the machine doesn't know what to do. A commonly

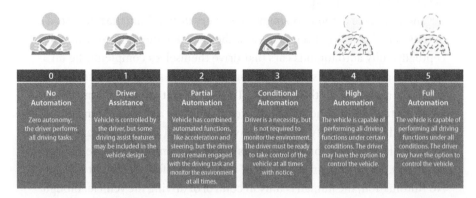

0	1	2	3	4	5
No Automation	**Driver Assistance**	**Partial Automation**	**Conditional Automation**	**High Automation**	**Full Automation**
Zero autonomy; the driver performs all driving tasks.	Vehicle is controlled by the driver, but some driving assist features may be included in the vehicle design.	Vehicle has combined automated functions, like acceleration and steering, but the driver must remain engaged with the driving task and monitor the environment at all times.	Driver is a necessity, but is not required to monitor the environment. The driver must be ready to take control of the vehicle at all times with notice.	The vehicle is capable of performing all driving functions under certain conditions. The driver may have the option to control the vehicle.	The vehicle is capable of performing all driving functions under all conditions. The driver may have the option to control the vehicle.

FIGURE 7.1. Description of Society of Automotive Engineers (SAE) Levels of Driving Automation. Source: SAE International and NHTSA.

SAE **J3016**™ LEVELS OF DRIVING AUTOMATION™

Learn more here: sae.org/standards/content/j3016_202104

Copyright © 2021 SAE International. The summary table may be freely copied and distributed AS-IS provided that SAE International is acknowledged as the source of the content.

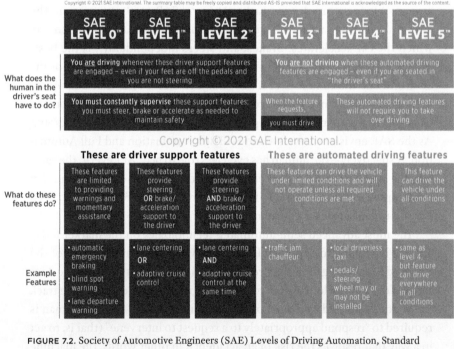

	SAE **LEVEL 0**™	SAE **LEVEL 1**™	SAE **LEVEL 2**™	SAE **LEVEL 3**™	SAE **LEVEL 4**™	SAE **LEVEL 5**™
What does the human in the driver's seat have to do?	You **are** driving whenever these driver support features are engaged – even if your feet are off the pedals and you are not steering			You **are not** driving when these automated driving features are engaged – even if you are seated in "the driver's seat"		
	You must constantly supervise these support features; you must steer, brake or accelerate as needed to maintain safety			When the feature requests, you must drive	These automated driving features will not require you to take over driving	
	These are driver support features			**These are automated driving features**		
What do these features do?	These features are limited to providing warnings and momentary assistance	These features provide steering OR brake/ acceleration support to the driver	These features provide steering AND brake/ acceleration support to the driver	These features can drive the vehicle under limited conditions and will not operate unless all required conditions are met		This feature can drive the vehicle under all conditions
Example Features	• automatic emergency braking • blind spot warning • lane departure warning	• lane centering OR • adaptive cruise control	• lane centering AND • adaptive cruise control at the same time	• traffic jam chauffeur	• local driverless taxi • pedals/ steering wheel may or may not be installed	• same as level 4, but feature can drive everywhere in all conditions

Copyright © 2021 SAE International.

FIGURE 7.2. Society of Automotive Engineers (SAE) Levels of Driving Automation, Standard J3016. Source: © SAE International from SAE J3016™, *Taxonomy and Definitions for Terms Related to Driving Automation Systems for On-Road Motor Vehicles* (2021-04-30), https://saemobilus.sae.org/content/J3016_202104/

used industry heuristic describes the progression of autonomy levels by the body part of the human driver that can be disengaged, under appropriate conditions, at different levels: feet-off (Level 1), hands-off (Level 2), eyes-off (Level 3), and brain-off (Level 4).

What would the handoff model mean for truckers? In theory, the truck would handle the bulk of the driving in good conditions, and the human trucker would take over in situations where the machine has trouble—say, in a construction zone or crowded intersection, or when visibility is poor. When the machine is in charge, the theory goes, the trucker might be freed up for other tasks: some commentators forecasted that "[i]f drivers are unshackled from the wheel, they could do order processing, inventory management, customer services and sales"[57] or other "higher skill activities"[58] while the robot drives. A former product manager at Otto—an autonomous trucking startup later acquired by Uber—had an even sunnier prediction: that when the truck drives itself, drivers could "nap and relax[,] . . . chat with family and friends, learn a second trade, or run a business."[59]

This vision is similar to the transformation of the bank teller's role after the advent of the ATM: the machine does the boring routine work, freeing up the human for more interesting or skill-matched pursuits. But it leaves open big questions about whether or how truckers would be paid for time in the cab while the truck drives itself—after all, if trucking companies are still paying big labor costs, are autonomous trucks worth the investment?—and also wouldn't necessarily address problems around overwork and fatigue.

There's another problem that's even more fundamental. In theory, and given due consideration to social and economic factors, the baton-passing model sounds like it could be an ideal arrangement. But in practice, baton-passing is incredibly—perhaps intractably—difficult to execute smoothly in situations like driving. Recall that the machine passes off responsibility to the human in the situations it finds most difficult: when conditions are unusual, when there is something in the environment it isn't equipped to contend with, when there's a mechanical malfunction or emergency. Those situations are very likely to be safety critical. One review of the scholarly literature found "a wealth of evidence" that automating some aspects of driving led to "an elevated rate of (near-) collisions in critical events as compared to manual driving. . . . Essentially, if the automation fails unexpectedly with very little time for the human to respond, then almost all drivers crash[.]"[60]

This problem is so severe because the time scale in which the baton is passed is miniscule: because of the nature of driving, a human is likely to have an extremely short window—perhaps only a fraction of a second[61]—in which to understand the machine's request to intervene, assess the environmental situation, and take control of the vehicle. This tiny time window is the reason why human drivers in semi-autonomous cars are warned that they must stay alert the entire time the car is driving. (Of course, in practice, humans ignore this warning, with important consequences—more on this later.) Despite the image of humans relaxing, napping, texting, eating, and being otherwise freed up from the requirements of driving, this image is patently unrealistic given the need for quick, safety-critical handoffs at current levels of automation. Audio and visual alarms can help humans know when a handoff is coming, but the immediacy of the need to take control means that humans must still pay constant attention.[62] However, a 2015 NHTSA study found that in some circumstances it could take humans a full *seventeen seconds* to regain control after a vehicle alerted them to do so—long beyond what would be required to avoid an accident.[63]

This problem may be further exacerbated in the future because our skills—if rarely used over time—tend to atrophy. The less often humans are called upon to drive a truck, pilot a plane, or otherwise do manually something which our robot coworkers have taken over day-to-day, the rustier we become at doing so. In 1983, engineering psychologist Lisanne Bainbridge wrote a prescient paper called "Ironies of Automation" that highlighted the paradox created when human workers become primarily "machine minders":

> [A human operator] may be expected to monitor that the automatic system is operating correctly, and if it is not he may be expected to call a more experienced operator or to take-over himself. . . . Unfortunately, physical skills deteriorate when they are not used, particularly the refinements of gain and timing. This means that a formerly experienced operator who has been monitoring an automated process may now be an inexperienced one. . . . When manual take-over is needed there is likely to be something wrong with the process, so that unusual actions will be needed to control it, and one can argue that the operator needs to be more rather than less skilled[.][64]

Skill atrophy leads to cognitive slowdown that can be crucial in a safety-critical situation; though the physical mechanics of how to do something may remain ingrained, humans do much worse with skills like performing mental calculations, recognizing abnormalities, navigating and visualizing how things relate to one another in space, and other cognitive tasks.[65] And subsequent generations of humans, who might much more rarely experience fully manual driving, will have even smaller, weaker skillsets to fall back on. A 2016 report from the Federal Aviation Administration raised the concern in the context of airplane automation—noting that human pilots may lack sufficient practice in operating a plane manually, and recommending that airlines provide more manual training opportunities to help pilots keep these skills sharp.[66]

Not only is it hard for humans to intervene when intervention is called for, it's cognitively unrealistic to expect humans to remain alert to the environment in case of emergencies—particularly as those emergencies become rarer, and as people begin to trust the technology more fully. This is what human factors researcher Peter Hancock calls the "hours of boredom and moments of terror" problem.[67] Humans are notoriously bad at staying attentive to monotonous situations in which there is only rarely something extremely important for them to notice and act upon. It's clear (particularly in the context of trucking) that work makes humans tired. But there's another kind of exhaustion, known as "passive fatigue,"[68] that comes from having *not enough* to do—what psychologist Stephen Casner and colleagues term "the tiring task of doing a little less."[69] In fact, passive fatigue can be even *more* dangerous in the driving context than active fatigue (that is, fatigue that comes from being overloaded with work tasks), in terms of its effects on reaction times—and, ultimately, likelihood of a crash.[70] As Hancock frames it: "if you build vehicles where drivers are rarely required to respond, then they will rarely respond when required."[71]

Studies of *vigilance decrement*—our tendency to let the mind wander from a boring task over time—were first conducted during World War II, when radar operators were charged with watching a display for hours to try to detect irregular blips. In 1948, psychologist Norman Mackworth developed a "clock test" in which study participants were given the mind-numbing task of monitoring the ticking hand of a blank-faced clock for two hours. When the clock hand jumped further than usual—the "signal"—participants were told to press a button. Mackworth's test found that people became markedly less likely (10 to 15 percent) to detect the signal accurately after thirty minutes—and this ability declined even more as the task continued.[72] Subsequent studies have confirmed and extended Mackworth's findings, finding that drivers in automated vehicles begin to show signs of tiredness and distraction after only ten minutes[73]—problems which get worse over longer drives.[74]

So it is difficult for humans to quickly understand and act on emergency situations, particularly as their skills degrade, and to monitor a machine with vigilance for long periods of time.[75] This creates a fundamental irony about automation and human labor, argued Bainbridge; in contrast to its labor-saving intent, "[b]y taking away the easy parts of his task, automation can make the difficult parts of the human operator's task more difficult."[76]

This irony creates severe problems for human/robot handoffs in autonomous cars and trucks. So long as humans have *some* duty to monitor the driving environment—which they do at Levels 2 and 3, the current state of the art—humans will almost inevitably do a poor job at accepting the baton from the machine. Does this mean there's no hope for safe autonomous vehicles? Not necessarily. If robots and humans make bad coworkers because of the weaknesses of the human, one solution might be to increase the level of automation even more, obviating the need for short-term handoffs to a human at all. This could create a second model of integration: *network coordination*.

Network Coordination: Divide and Conquer

Our discussion of handoffs and baton-passing has focused on the nitty-gritty of specific driving tasks: who (or what) will take the wheel in what conditions, who (or what) is responsible for recognizing a pedestrian and looking out for the unexpected. But another way to think about the division of labor between humans and machines is as a matter of more systemic work-sharing. Rather than a focus on in-the-moment driving, we might think about humans and machines as sharing truck-driving work in a broader way: by dividing up responsibilities over the driving *route*. We've been thinking about the work of truck driving as a set of small, often simultaneous driving tasks: change lanes, hit the brakes, watch for road obstacles. We could instead think about it as a series of predictable *segments*: travel down the interstate, exit the highway and take local roads, steer around the receiver's docks. In this model, humans and robots still share the labor of trucking work, but take turns being wholly responsible for driving—much as you and a friend might take turns driving on a road trip—with temporally and geographically predictable points of transition between the two of them. Some truckers already do this when they "drive team," taking turns driving (often while one driver sleeps). If we think of human/robot teams working together in tandem over these segments, a second model of integration emerges: *network coordination*. As we'll see, several trucking technology firms have set their sights on this sort of model.

But wait, you might think. The reason for autonomous cars to hand off control to humans is that they *aren't* fully capable of driving themselves—they can't negotiate unexpected obstacles well, they lack humans' tacit knowledge, they can fail catastrophically in new and complex situations. If this happens, how can we envision giving a machine total control over an entire portion of the route, without a human driver being expected to step in?

Part of the answer is that the difficulties autonomous vehicles encounter are "lumpy"—they're much more likely to occur in some route segments than others. Though they're far from perfect in any setting, autonomous vehicles perform much better on highways than on city streets: speeds are more constant, there are fewer intersections and unexpected obstacles, and contexts are generally more predictable and easier for a machine to negotiate. Things get much more complicated at the endpoints, when trucks leave the highways and venture into cities and towns to pick up or drop off loads. And when a truck arrives at a terminal, it doesn't just drop its load immediately and take off. A trucker might spend hours at a terminal making "yard moves"—queuing to be loaded or unloaded, backing the truck into the right bay, and following the directions of the customer. Some truckers load and unload freight themselves; others coordinate with the customer's unloading crew (or with "lumpers," third parties who unload the delivery on behalf of the customer). All of this requires irregular driving in response to immediate human direction, sometimes in large lots without lanes or traffic markings—and is nearly impossible for a machine to do on its own. (As a point of comparison, think of how planes taxi around at airports—despite the widespread use of autopilot in the air, there's little chance that airport taxiing will be automated anytime soon.) So, a natural division of labor in trucking might be that advanced autonomous trucks drive themselves over the long haul, and humans take the wheel for the endpoints—what's often called the "last mile" in transportation and logistics.

But still, you might think again: even over the long haul we've got problems. The state of the art in autonomous vehicles is still Level 2, which requires constant human monitoring, and brings with it all the intractable handoff problems we've discussed. So how is this a feasible model?

It isn't—yet. But some autonomous vehicle technology companies think the human/machine coordination challenges at Autonomy Levels 2 and 3 are so difficult and intractable that they are essentially attempting to "skip" those levels, focusing their attention on developing Level 4 autonomy instead—which allows for full autonomy, no human required, within particular conditions (for example, highway driving within a prespecified area, under only certain weather conditions, etc.). They're doing this both for technical reasons and economic ones. In trucking, Level 3 offers little economic benefit: if a human trucker must stay in the driver's seat monitoring the road at all times (and growing more fatigued the longer he sits there), it's not obvious why the firm would invest substantially in the technology. But if Level 4 enables the truck to drive *without* the driver's constant attention (and for longer time periods—since robots don't get tired), the prospect is much more economically viable.

Is it possible to "skip" intermediate levels of automation and go straight to Level 4? Many carmakers are trying to do so. Volvo announced plans to do so, claiming that it believes Level 3 driving to be inherently unsafe.[77] Ford also announced that it would "abandon[] a stepping-stone approach" and push straight to full autonomy.[78] Daimler Trucks made a similar move, justifying the choice on the belief that "Level 4 is safer than Level 3."[79] Google's autonomous car division (now Waymo) raised eyebrows back in 2014 when it released a self-driving prototype with no steering wheel or brake pedal.[80] These experimental cars, intended for Level 4 autonomous driving, were so different from what people were used to that they were hard to stomach—but the company justified its vision based on what it had learned about intractable difficulties with the handoff at lower levels of autonomy, and they seem to have been justified in doing so.

Uber, more popularly known for its ridesharing technologies, has tried to capitalize on this version of coordinated human/machine labor. In 2018, Uber announced that it envisions an autonomous truck network, connected by local (and presumably Uber-owned) "hubs" throughout the country. In this vision, customers would contract with Uber for the delivery of goods; autonomous trucks would run the long hauls between hubs, and human truckers would pick up and pilot the trucks from the hubs to delivery. Uber described this model as "a future where truck drivers and self-driving trucks *work together*,"[81] handing off loads to one another at the transfer hubs. This model exploits Uber's existing expertise with respect to distributing rides (or hauls) among drivers in a network. (In a separate but related effort, the company rolled out Uber Freight—an app that matches independent truckers with nearby available loads, replacing the load boards and brokers that typically orchestrate such matches—much like the driver/passenger matching that Uber specializes in for rideshare.)

Uber described its network coordination model in sunny terms; it hoped that its system would "lead to a more optimistic outcome"[82] than the mass job losses many feared might result from the new technology. Uber imagined instead that many trucking jobs would shift primarily from long-haul to local hauls. They further claimed that the increased efficiency of its model (self-driving trucks can drive all night and day) could boost the economy generally, increasing consumer demand for goods (and accordingly, more trucking jobs to move those goods).

Uber's assumption that the increased logistics efficiency wrought by automation would trickle down into consumer demand and more jobs is a big assumption—and not one shared by all economists. But a model like this one *could* make human truckers' work better. Local drivers aren't on the road for weeks at a time, far from their families and communities—they go home every night. And importantly, local trucking would likely be safer and healthier for human truckers: by cutting out the long stretches of highway driving, the model could greatly reduce the driver fatigue that plagues the industry now.

But there's one problem: recall that truckers are paid by the mile. The great majority of miles driven (and thus money earned) takes place on the highway—not on traffic-packed local roads, not while maneuvering around at a terminal at the end of a haul. The parts of the job that Uber proposed to automate are precisely the parts that make up the lion's share of a trucker's wages! They make little or nothing for the tasks to which they'd be exclusively relegated under Uber's model. Under the current dominant pay scheme, Uber would be hard-pressed to convince human drivers to focus entirely on the least profitable aspects of their work.

The only way the network coordination model is a viable option is if the pay structure of trucking adjusts with it. Truckers have argued for pay reform in the industry for decades, but have lacked the political capital to make change. Uber, on the other hand, is an immensely powerful company that would be much more likely to have the power to transform the industry—much as it has reshaped the taxi industry, and repeatedly lobbied powerfully for its own interests in the context of labor and employment law (primarily around the status of gig workers as employees or independent contractors). What's more, Uber began publicizing its network plans in 2017, shortly after it began to come under an unwelcome public spotlight for its questionable corporate culture and labor practices—a time when it might have welcomed a worker-friendly "win." So Uber's proposal initially seemed as though it might be an unholy alliance that could actually *help* improve truckers' lots and create a viable way forward for humans and machines to work together.[83]

In July 2018—only months after Uber rolled out announcements about its hub-and-spoke model, to much fanfare—the company abruptly shuttered its autonomous truck division, saying it had decided to focus its energies exclusively on self-driving passenger cars.[84] In December 2020, it ended its efforts with respect to self-driving passenger cars, too.[85] It's not entirely clear what precipitated these shifts, though many in the industry chalked it up, at least in part, to messy legal troubles between Uber and Waymo (Google's self-driving vehicle group); Uber had allegedly poached trade secrets from Waymo in a much-publicized controversy that cost Uber about $250 million in settlement costs and landed engineer Anthony Levandowski an eighteen-month prison sentence.[86] (Levandowski was later pardoned by Donald Trump.) Other commentators chalked up Uber's self-driving woes to the technical challenges of autonomous vehicle production—not least, the intractability of the handoff problem.

With Uber's shift away from autonomous trucks went any serious hope of achieving the network coordination model anytime soon. The project would have been tremendously ambitious, even for Uber—it necessarily involves substantial regulatory and organizational change, as well as colossal infrastructure and logistics costs to build transfer hubs all over the country and to distribute loads efficiently at the hubs. Uber was one of the only companies that might have pulled it off in the near term.

A variation on network coordination could involve allowing truck drivers to take the wheel *remotely* for the "last mile" of operation. Starsky Robotics, founded with significant venture capital investment in 2016, developed a "teleoperation" system in which trucks drove themselves to a certain point, and human drivers subbed in remotely from the highway exit to the terminal[87]—as if they were playing a video game or operating a drone. In theory, such a system could allow a single driver to pilot dozens of vehicles a day, for short periods of time, all over the country—and still return home each night. (As one remote trucking executive framed it: "Think about the mom who is home driving a truck. She can drive multiple assets and never leave her kids.") Some refer to this as a "call-center" model in which the robot calls into a human phone bank for support or handoff at predetermined points in the route.[88] But it isn't obvious that a model like this is sustainable, either. For one, the handoff problems seem likely to be only exacerbated by distance.

And there are other problems unique to the model: Ford shut down its system testing a similar idea after the vehicles repeatedly lost their cell signal so that human operators couldn't see the video feed.[89] Starsky Robotics closed its doors in 2020; in a valedictory blog post, its chief executive chalked up the company's closure in large part to the assessment that "supervised machine learning doesn't live up to the hype" in terms of operational capability in autonomous trucking.[90]

Human/Machine Hybridization: The Rise of the RoboTrucker

The future of trucking might someday look like these baton-passing or network-coordination models of shared labor. But right now, human/machine interaction in trucking looks very different. What we see happening in trucking now involves a much less discrete parceling-out of functions between humans and machines. Instead, truckers' physical bodies and intelligent systems are being *integrated into one another.*

The idea of direct integration between human bodies and computers has historical roots. The idea of the *cyborg*—a human/robot hybrid, or *cybernetic org*anism—was first proposed by scientists Manfred Clynes and Nathan Kline in 1960.[91] Clynes and Kline were interested in adapting human bodies for living in space for long periods of time. They thought there were essentially two ways to do this: humans could bring equipment to space that allowed their bodies to function more or less as they would in a "normal" (Earth-bound) environment—or they could *adapt* their bodies to more naturally function in the environment of space. Clynes and Kline were far more excited by the second option—the cyborg—in which the human body is modified to better fit the environs in which it operates.

A key aspect of the cyborg ideal is that its machine capabilities are understood as an *enhancement* of the normal human body. Becoming part-human, part-machine is a way of becoming superhuman: machines extend a person's natural capabilities and leave the person "free to explore, to create, to think, and to feel."[92] Clynes and Kline wrote optimistically that "adapting man to his environment, rather than vice versa, will not only mark a significant step forward in man's scientific progress, but may well provide a new and larger dimension for man's spirit as well."[93]

Take a look at the image below, from a 1964 article in the *Miami News*[94] discussing the prospect of sending cyborgs into space:

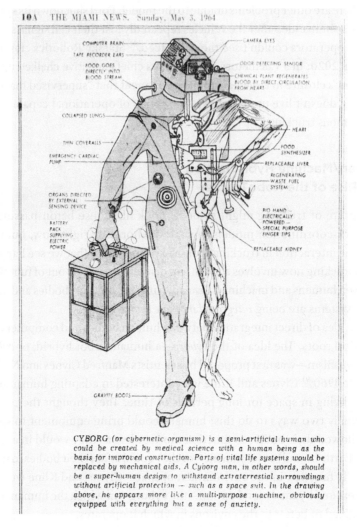

COMPUTER BRAIN

TAPE RECORDER EARS

FOOD GOES DIRECTLY INTO BLOOD STREAM

COLLAPSED LUNGS

THIN COVERALLS

EMERGENCY CARDIAC PUMP

ORGANS DIRECTED BY EXTERNAL SENSING DEVICE

BATTERY PACK SUPPLYING ELECTRIC POWER

GRAVITY BOOTS

CAMERA EYES

ODOR DETECTING SENSOR

CHEMICAL PLANT REGENERATES BLOOD BY DIRECT CIRCULATION FROM HEART

HEART

FOOD SYNTHESIZER

REPLACEABLE LIVER

REGENERATING WASTE FUEL SYSTEM

RIO HAND — ELECTRICALLY POWERED — SPECIAL PURPOSE FINGER TIPS

REPLACEABLE KIDNEY

CYBORG (or cybernetic organism) is a semi-artificial human who could be created by medical science with a human being as the basis for improved construction. Parts of vital life systems would be replaced by mechanical aids. A Cyborg man, in other words, should be a super-human design to withstand extraterrestial surroundings without artificial protection — such as a space suit. In the drawing above, he appears more like a multi-purpose machine, obviously equipped with everything but a sense of anxiety.

FIGURE 7.3. An imagined cyborg. *Source*: Toby Freedman, "'Robot Man' Not Space Solution," *Miami News*, May 3, 1964, p. 10, archived at https://perma.cc /JS9T-3GMK.

The "machine-man" pictured here sports "tape recorder ears," "camera eyes," and "special purpose finger tips." He is "equipped with everything but a sense of anxiety" as he traverses the new environment in which he finds himself, and is made more capable than ever by the technology that seamlessly integrates with his body.

But the idea of the human/machine hybrid doesn't always connote freedom and potential.[95] In the workplace, the integration of human bodies and technology can instead be seen as a way for managers to keep ever-closer tabs on workers. At the most extreme, a few companies have begun offering workers microchip implants—just like the chips we commonly implant into our pets. These chips, usually implanted into the hand, use near-field communication technology to be "read" by sensors throughout the workplace: they give workers access to restricted areas, let them use copiers and printers, and can be scanned to purchase food in the cafeteria.[96]

Though implants are still an exceptional case, a large number of employers use wearable technologies to monitor workers. Neck-worn "smart" ID badges give employers insight into precisely where workers are in the office, for how long, and with whom they interact, and let managers measure and incentivize behaviors that the firm considers productive.[97] Amazon has patented location- and motion-tracking wristbands for use in its warehouses that can track how long it takes for a worker to pull an item from the shelf and whether the worker takes a too-lengthy bathroom break, and that vibrate when the worker makes an error.[98]

These sorts of wearables are generally found *on* rather than *in* the body, but often draw on data about bodily systems and behaviors. Workplace wellness programs may collect biometric data from wearable devices like Fitbits to incentivize healthy behaviors (like physical activity) through insurance discounts. Face recognition technologies and other biometrics are used for clocking in and out of hourly jobs and for measuring workers' moods.[99] Another Amazon patent application for an "augmented reality user interface facilitating fulfillment"[100] describes a pair of goggles to be worn by warehouse workers that projects instructions within the worker's field of view, like directions to a particular location within a warehouse and information about items to be packed. The goggles double as a tool for fine-grained monitoring of what the worker is doing, the speed at which they move, the direction in which they are looking, and the like.

The human/machine hybrid being realized in these contemporary labor scenarios seems a far cry from the enhanced, worry-free cyborg Clynes and Kline imagined. In her 2008 book *The Culture of Soft Work*, Heather Hicks describes the cyborg as a cultural icon of modern work: when "work activities coded in machine parts suffuse the human body," she writes, "the results are humans not liberated, but under control."[101] Microchips and augmented-reality goggles marry a worker's unique bodily and cognitive capabilities with those of the machine, turning them into a "super-worker" of sorts—but also operate as tools through which the worker can be more closely supervised and managed.

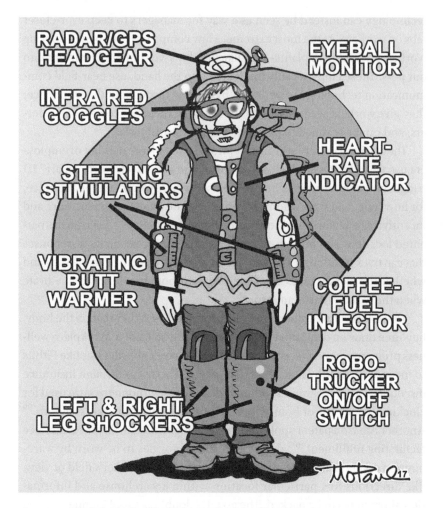

RADAR/GPS HEADGEAR

EYEBALL MONITOR

INFRA RED GOGGLES

HEART-RATE INDICATOR

STEERING STIMULATORS

VIBRATING BUTT WARMER

COFFEE-FUEL INJECTOR

ROBO-TRUCKER ON/OFF SWITCH

LEFT & RIGHT LEG SHOCKERS

FIGURE 7.4. The RoboTrucker. Credit: Mo Paul and *Land Line* Magazine.

Compare the cyborg man from the *Miami News* illustration to the cartoon in figure 7.4, which appeared in a 2017 blog post published by *Land Line*, a truckers' trade magazine.[102]

The article accompanying the illustration laments the rise of the "RoboTrucker": the driver burdened by a proliferation of gadgets that aim to keep him awake and alert. The technologies shown in the illustration are exaggerated for effect, but only just: a number of products on the market do more or less what the cartoon depicts. These tools generally are marketed as *fatigue detection* systems or *lone worker monitoring* devices, aimed at giving managers remote insight into how tired a trucker's body is, and how fit he is to drive at that moment.

Generally speaking, there are two kinds of technologies that turn truckers into RoboTruckers. The first are wearables, which monitor elements of the trucker's internal bodily state and use them as metrics for management. For example:

- The SmartCap is a baseball cap (also available as a headband) that detects fatigue by monitoring a driver's brainwaves (essentially doing a constant EEG). Rear View Safety and Ford's Safe Cap are similar systems. Systems like these can be configured to send an alert to a fleet manager or a family member, to flash lights in drivers' eyes, to sound alarms, or to jolt the wearer back to alertness with vibrations. Similar devices are used by high-speed train conductors in China to monitor workers' stress levels and emotional spikes.[103]
- Optalert, an Australian company, manufactures a pair of glasses that monitors the speed and duration of a trucker's blinks in order to give him a real-time fatigue score.
- Maven Machines' Co-Pilot Headset detects head movement that suggests the driver is distracted (for example, looking down at a phone) or tired (for example, failing to check his side-view mirrors regularly).[104]
- Wrist-worn Actigraph systems both monitor *and predict* fatigue rates over time. The technology, initially developed by an Army research lab, blends biometric data about a trucker's alertness with other data (like start time) to forecast how long he can drive before becoming too tired.[105]

A number of other wearable devices are under development. For example, Steer, another wrist-based wearable being developed by a Latvian firm, measures heart rate and skin conductivity. It vibrates and flashes lights if it begins to detect signs of fatigue and delivers a "gentle electric shock" to the driver if fatigue continues.[106] Mercedes has prototyped a vest to monitor a trucker's heart rate; the system can stop the truck if it senses the trucker is having a heart attack.

The second set of technologies are cameras pointed at the driver designed to detect his level of fatigue, often by monitoring his eyelids to track his gaze and look for signs of "microsleep." Seeing Machines is one of several companies that market driver-facing cameras that use computer vision to monitor a driver's eyelids and head position for signs of fatigue or inattention. If the

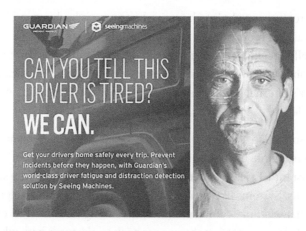

FIGURE 7.5. From Seeing Machines' marketing materials.

driver's eyes close or look away from the road for too long, it sounds an alarm and sends a video to his boss[107]—and can also causes the driver's seat to vibrate in order to "goose" him back into attention.[108] Another driver-facing camera vendor, Netradyne, uses deep learning and data from driver- and road-facing cameras to generate scores for drivers based on their safe and unsafe driving behaviors.

Some industry insiders believe that it's only a matter of time before trucker wearables and driver-facing camera systems become standard—or even legally required, following the path of the ELD before them. Firms will value these systems not only for fatigue detection, but as a source of evidence in accident investigations,[109] and even as a means of tying drivers' performance to compensation (for example, by giving drivers bonuses for not triggering fatigue alarms).[110] If this sounds familiar, it should; the same managerial dynamics that supported the adoption of the ELD seem very likely to be similarly tied to biometric and camera systems in the near future. And just as previous managerial tools were baked into the ELD's hardware, fatigue detection tools seem destined to become part and parcel of the very same systems. For example, Blue Tree Systems' ELD offers managers the ability to score drivers on a number of driving behaviors, and also tracks the time between a driver taking his foot off the accelerator and pushing it onto the brake—a measure called "anticipation"—which can operate as a proxy for fatigue (as tired drivers tend to make these moves more suddenly).[111] Driver-i's machine vision capabilities not only monitor drowsiness, but check to ensure the driver's seatbelt is buckled.[112] There are also early indications that such systems might be of interest for insurance purposes;[113] one carrier's safety director said he expects a mandate for use of fatigue monitoring "not from the feds[,] but from the underwriters."[114]

Others raise the possibility that fatigue-detection systems might upend the hours-of-service regulations altogether[115]—after all, why not know *precisely* how tired a driver is, rather than making a blanket rule that assumes everyone has the same capabilities? This prospect raises a number of interesting issues. Might the trade-off for clumsy hours restrictions be an intrusive invasion of truckers' bodily privacy? To what extent do the potential benefits of a more personally tailored regulation—not to mention the public safety benefits of less tired truckers—warrant requiring truckers' intimate bodily data to be collected by their employers, much less the government? So far, these events have not come to pass. In 2020, Pronto.ai—which sells a Level 2 autonomous truck system that includes camera- and sensor-based driver monitoring, fatigue detection, and other capabilities—filed a petition with the FMCSA requesting that drivers using its system be exempted from the hours-of-service rules. It argued that the exemption would provide "an operational 'carrot' to encourage adoption of Level 2 [autonomous systems]" that could "help overcome driver reluctance to accept and use critical safety technologies."[116] (FMCSA later denied the petition, largely because of a dearth of data and no defined safety metrics that would permit evaluation of Pronto.ai's safety claims.)

It will likely be a while before we do away with the hours-of-service regulations altogether, but the government is beginning to dip its toe into using AI-driven technologies for enforcement purposes. As far back as 2013, the FMCSA's Medical Review Board urged the regulator to consider requiring cameras for monitoring driver performance.[117] AAA has also recommended requiring driver-facing cameras as a safety measure.[118] As another example, a program at the US-Canada border has begun to use facial recognition software to scan drivers' faces as a "fast track" through customs enforcement; by scanning the driver's face, Customs can know instantaneously whether the trucker "is up-to-date on his paperwork and border fees[.]"[119]

The *Land Line* article acknowledges that these sorts of technologies can help truckers in some contexts—giving them needed directions, helping them stay safer and better connected. But it raises concerns about the slippery slope such technologies create, arguing satirically:

> Why can't we produce boots for the right foot that "shock" the driver if he or she applies too much or too little pressure on the gas pedal? And how

about some gloves that create a small "burn" if the driver's hands don't remain in the correct position on the steering wheel? While we're at it, can we get a helmet that tracks a driver's thoughts, and produces a "stinging" sensation if he or she begins thinking about anything other than driving?[120]

As this passage makes clear, from the truckers' point of view, there's something viscerally offensive about the micromanagement enabled by these technologies. The privacy intrusions brought about by the introduction of the ELD pale in comparison. *This* is the felt reality of AI in trucking labor now: using AI to address human "weakness" through constant, intimate, visceral monitoring. There's an enormous distance between the narrative of displacement that characterizes most public discussion of AI's effects on truckers and how these effects are actually being experienced through these technologies. The threat of displacement is a real one, particularly to truckers' economic livelihood—but driverless trucks are not yet borne out by common experience, and drivers are also not yet handing off a baton to or splitting routes with a robot coworker. Truckers' encounters with automation and artificial intelligence have not yet supplanted them.

Instead, technologies like the ones we've discussed above represent a distinct and simultaneous threat: a threat of *compelled hybridization*, an intimate invasion into their work and bodies. AI in trucking today doesn't kick you out of the cab; it texts your boss and your wife, flashes lights in your eyes, and gooses your backside. Though truckers are, so far, still in the cab, intelligent systems are beginning to occupy these spaces as well—in the process, turning worker and machine into an uneasy, confrontational whole.

Automation and Surveillance as Complements

Telling the story of how AI is finding its way into trucking helps us to understand workplace *automation* and workplace *surveillance* as complements, not substitutes. Trucks will be "driven" by humans for a long time to come, though what "driving" means is likely to change—and we must consider what these changes mean for the humans tasked with doing the job. They may mean that the job is less risky, less boring, and better for all involved—or they may mean that the worker is monitored more intimately and more intrusively by the technology he works alongside.

A skeptical reader might chalk all this up to growing pains. There are always winners and losers in technological change; some people's jobs disappear, some improve, and others get worse. Such is the price of progress. And perhaps all these intrusive technologies prove is how much we *need* fully autonomous systems in the first place. Humans are faulty in ways that robots are not: humans get tired, they get distracted, they have heart attacks and strokes. The faster we get humans and their imperfections out of the loop, the better off everyone might be.

It's true that humans have faults—but automating them away isn't a solution. There will always be a role for humans in (so-called) autonomous systems—in their design, operation, maintenance, use, and oversight. Those roles may look different than they did before, but the notion that humans *can* be eliminated from systems is fundamentally false.

This tells us something important about the relationship between automation and surveillance. A naïve reading of the situation might suggest that automation obviates the need for surveillance. But recall that even the most sophisticated autonomous vehicle systems on the market today require humans to maintain attention and be at the ready to step in. And recall that humans are *extremely bad* at maintaining such attention—and will look for creative ways around doing so. So how do we impel humans to fulfill their roles? By monitoring them, and nagging them, and imposing legal liability on them if they fail. Despite marketing rhetoric of humans kicking back and napping while their cars autonomously zip them around town, the consumer market for autonomous vehicles is rife with surveillance—in fact, the same sorts of technologies to which the RoboTrucker is subjected.

In a 1970 article called "Big Brother as Driver: New Demands and Problems for the Man at the Wheel," robotics scholar Thomas Sheridan foresaw that "in very few years semi-automation will begin to encroach upon the autonomy of the automobile driver."[121] Sheridan's prophecy is being fulfilled in today's consumer vehicles. There are abundant examples.[122] GM's Driver Attention System engages when its hands-free "Super Cruise" mode is enabled; the system's camera monitors the driver's head and eyes, and uses flashing lights, voice alerts, and seat vibrations if it detects that the driver is inattentive. (It can eventually stop the car and contact emergency services if the driver fails to take control.) The system integrates software from Seeing Machines, which also makes the trucker eye-monitoring software discussed above. Audi uses face recognition and steering wheel sensors to monitor drivers—if the driver's attention drifts, it escalates warnings, and will eventually stop the car if the driver doesn't respond; Toyota markets a similar system.[123] Volvo's cars will suggest the driver take a coffee break if the car's movements become erratic. The emotion analysis company Affectiva has also gotten into the driver monitoring business, incorporating data like the sound of the driver's voice, and inferring drivers' emotions from facial expressions, to predict distraction and drowsiness. The German automotive company ZF's Interior Observation System monitors not only the driver, but the location and position of other passengers in the vehicle—inferring, for example, if a child is left unattended in the car.[124]

As driver monitoring systems become more ubiquitous, they sometimes engage in an arms race with drivers who find the systems annoying and attempt to thwart them. Since self-driving marketing rhetoric promises that drivers will be "freed from an imagined burden of driving," as science and technology studies scholars Chris Tennant and Jack Stilgoe describe it[125]—despite the fact that the technology doesn't actually allow for this—it's no surprise that consumers who buy semi-autonomous cars often bristle at beeps and dings that require them to pay constant attention to the road and keep their hands on the wheel. And it's also not surprising (recalling our discussion of resistance practices in chapter 6) that these consumers seek out creative means to avoid such control and trick driver monitoring systems, despite the fact that it is extremely unsafe to do so. Tesla has been roundly criticized for how easy it has been to thwart Autopilot. One can easily find YouTube videos and Reddit threads in which Tesla owners vouch for various "hacks" to convince their cars that their hands are on the steering wheel (often involving wedging or taping various heavy objects to the wheel). In 2018, NHTSA ordered a company to stop selling an aftermarket product called the "Autopilot Buddy," a weighted piece of plastic marketed as a "Tesla autopilot nag reduction device."[126] But this may have been a drop in the bucket: a 2021 *Car and Driver* investigation tested driver monitoring systems in seventeen different vehicles, and found that *all* could be readily tricked by a motivated driver. Most could be fooled by draping an ankle weight across the steering wheel; eye-tracking software could be tricked "with a pair of gag glasses emblazoned with eyeballs."[127]

Even if not foolproof, regulators are very likely to encourage or require more stringent driver monitoring systems in the future. After a fatal 2016 crash in which a Tesla driver apparently ignored Autopilot's visual and audible reminders to put his hands on the wheel (but the car continued to drive in Autopilot mode anyway), the National Transportation Safety Board found that Tesla's driver monitoring system was "not an effective method of ensuring driver engagement" and suggested other avenues, including eye-tracking cameras.[128] (Tesla initially ignored these recommendations, along with similar appeals from its own engineers—but began building driver monitoring cameras into its AV models in 2021.[129]) European regulators will require driver monitoring systems in all new vehicles sold beginning in 2026; similar legislation has been introduced in the United States.[130]

My aim here is not to suggest that some degree of driver monitoring is not acceptable or necessary for semi-autonomous systems to operate safely. Rather, it is to recognize that surveillance is necessarily part and parcel of autonomous systems because *people are still integral to those systems*, and are likely to remain so. Therefore, discussions about the desirability of autonomous systems must necessarily also be discussions about how much surveillance we are willing to stomach to make these systems work.

———

Back to the truckers. Remember, from our discussion in chapter 2, that the culture of trucking is premised on an idealized notion of freedom and autonomy. Yet the open road is in fact much *less* free than it might seem, as truckers face a tangle of regulations and controls as they complete their daily work. A similar contradiction underlies AI technologies in trucking. The narrative is similar: that autonomous vehicles will free drivers up for rest and relaxation; that they'll free workers up for new and more interesting tasks; that they'll free the trucking industry from worrying about unreliable, fallible drivers and labor shortages. Yet none of these narratives holds up under scrutiny or accurately describes the effects of AI in trucking.

In reality, truckers face a double threat from AI technologies. On one hand, concerns about worker displacement are real and important—even if displacement occurs more slowly and gradually than some forecast. On the other, integration of AI into truckers' work is experienced as an intimate biological incursion into their work and bodies, only deepening the monitoring capabilities we've discussed throughout this book. When we evaluate these technologies or forecast their future use, we must account for each of these disparate effects simultaneously—and not only in terms of the number of jobs at stake, but also in terms of the quality and dignity of the work that remains.

8

Technology, Enforcement, and Apparent Order

In *Against Security*, his critique of the post-9/11 security apparatus, sociologist Harvey Molotch draws a useful distinction between two different views of how the social world works: its apparent order and its actual order. Molotch argues that all too often, we impose top-down solutions to social problems based on how we *think* the world functions (or at least, how we think it *should* function)—while disregarding, deliberately or out of ignorance, how the world *actually* functions. In other words, we impose apparent order to the detriment of actual order. As Molotch explains:

> [It is b]est to replicate or at least build on people's complexly tacit and mutual means of doing the world rather than trying to invent a world that does not exist. Too often the temptation is to reform through rules and official procedures that can sidestep those usual ways of doing things. But mundane life . . . should be interfered with only on pain of screwing things up in a big way. To create apparent order, you kill the actual order. . . . [Y]ou efface the tacit mechanisms and social work-arounds that people use to get things done[.][1]

Molotch's concern manifests in spades when we use digital systems to enforce rules. In the trucking industry, the imposition of a digital enforcement regime wholly fails to account for numerous elements of the actual order of the industry—the political, logistical, and economic realities with which the industry and its workers must contend. These include: the fact that truckers' economic well-being has long depended on widespread noncompliance with federal regulations; the logistical realities of unpaid detention time; the dearth of safe available truck parking and the lack of political will to build more; the unpredictable contingencies of the public highway, from weather to accidents to traffic congestion; the work pressures, sometimes coercive, from employers; cultural norms around machismo and stamina, which manifest themselves both in economic provision for family and in resistance to authority; and the occupational pride and deep-seated identity founded in a long tradition of workplace autonomy, just to name a few.

To be entirely clear: by no means do I intend to suggest that the actual order in today's trucking industry is economically equitable, socially just, or without need of serious reform—far from it! The actual order of trucking, as I've described it in this book, exploits workers and risks public safety. It's no wonder there is so much driver turnover in the industry. An incredible amount of economic activity rests precariously on a fundamentally unsustainable system in which workers are uncompensated for far too much of their dangerous and difficult labor. Perhaps surprisingly, I also do not intend to argue here that there is no place whatsoever for ELDs in trucking. ELDs—*if integrated alongside other meaningful economic reforms*—might well be one component of a healthy, reorganized trucking industry.

What I am arguing is that, by using digital surveillance to enforce rules, we focus our attention on an apparent order *that allows us to ignore* the real problems in the industry, as well as their deeper economic, social, and political causes. Under the apparent order envisioned by the ELD, the fundamental problem in trucking is that truckers cannot be trusted to reliably report how much they work, and the solution to that problem is to make it more difficult for them to fudge the numbers. But under the actual order, the problem in trucking is that drivers are incentivized to work themselves well beyond healthy limits—sometimes to death. The ELD doesn't solve this problem, or even attempt to do so. It doesn't change the fundamentals of the industry—its pay structure, its uncompensated time, its danger, its lack of worker protections. At best, it prevents some of the very worst-of-the-worst carriers from pushing drivers too far—but as we have seen, it remains both possible (and sometimes encouraged!) for drivers to exploit the limitations of electronic monitoring. It's a failure to recognize the actual order that leads drivers, under the gun of companies' profit motives, to compensate for the ELD by speeding to reach their destinations on time, making them (and all of us) less safe on the highway. It's a failure to recognize the actual order that leads technologists and trucking firms to embrace the promise of autonomous trucks—even though these technologies are nowhere near safe enough to deploy on their own, and end up leading to halfway solutions that imperil safety and intrude on drivers' bodies.

There *are* solutions to these problems. But they're not technology solutions. Technology fixes on their own are, at best, Band-Aids—they're superficial ways of covering up a problem, and don't address root causes. More meaningful reform in trucking would require a ground-up rethinking of how the industry is structured economically, in order to make trucking a decent job once more. So long as trucking is treated as a job that churns through workers, these problems won't be solved. Transport economist Michael Belzer argues that drivers leave the industry "when they realize the big promises aren't real"—when they learn that they'll work much longer hours than it appears on paper, when they come to see the open road as a sweatshop on wheels, when they realize that carriers are often less invested in their long-term success than they are in finding "fresh meat" to cycle into the job. As Belzer puts it, truckers "quit the industry because it spits them out."[2]

But if we *paid truckers for their work*, rather than only for the miles they drive—for the *actual* number of hours they work, including time they are waiting to be loaded and unloaded at terminals, time they are inspecting their trucks and freight, time they are taking the rest breaks that their bodies need to drive safely—drivers would be far less incentivized to drive unsafely or when fatigued. Remember that truckers are not covered by the protections of the Fair Labor Standards Act, which prevents them from being fairly compensated for time that they work—they are not entitled to a minimum wage or to overtime. Pay for time worked, not just miles driven, is especially essential as the specter of the autonomous truck looms, which will likely be able to drive highway miles (where truckers currently make the vast majority of their money) before local miles.[3]

Could collective bargaining help turn the tables? Maybe. But it will be difficult to create a culture of unionization where very little exists. Recall that only a negligible number of drivers in the industry today are union members. Legal reform to restructure the way drivers are paid, and to give them the protections due to other workers, is an essential step.[4]

What Role for Technology?

It's very attractive for decision-makers in power to seek out technological solutions to social and economic problems. They often seem easier to implement, they require buy-in from fewer stakeholders, and—most importantly—they often don't challenge existing institutional power structures nearly to the degree that other reforms might. Technology rarely, if ever, can solve social and economic problems on its own. But can it have a role in doing so? I believe the answer is yes. Digital data collection can be marshalled as *part* of a toolkit for addressing complex problems. The key is to avoid viewing technology as a solutionist panacea, and to recognize what roles it is best positioned to fulfill—and what roles it isn't.[5]

One thing technology can do well is help us to diagnose social and economic problems with greater precision.[6] Data collection can provide important, legible evidence of the shape and scope of a problem. A clear example from trucking is the secondary use of ELD data to better quantify the detention time problem. Recall from chapter 4 that ELD data, aggregated across hundreds of thousands of trucks, can be used to give truckers insights into how long waits are likely to be at different shippers and receivers, which can help them decide whether to accept a load. Though truckers have long known that detention time is a tremendous operational problem and source of fatigue, they have not previously had visibility into comparative average wait times across the country. And recall also that these data are also being used to advocate for truckers in Washington by tracing calculable safety outcomes to detention time. This, too, puts the problem of detention time into a form of knowledge that is much more legible to FMCSA, and that might provide a basis of evidence for addressing the issue through regulation.

Note that in both of these cases, the entity that ELD data are being marshalled to reform is no longer the trucker and his timekeeping practices: rather, it's shippers and receivers (in the case of wait time visibility) and federal policies themselves (in the case of FMCSA advocacy). The tables are turned. This is an example of using data analysis to "study up." As Chelsea Barabas and her colleagues describe it, "studying up"[7] involves refocusing data analysis not on the most marginalized actors in a system, but on more powerful actions and institutions; doing so might reorient data collection from a tool for control into a means of creating more systemic accountability and meaningful reform.

Of course, we should be cautious in our enthusiasm about the capacity for these data to facilitate systemic change in trucking. Even if we couldn't put hard numbers on average wait times before, detention time has been a known problem in the industry for years; assigning a number to it does not magically create the political will to address the issue. As artist and researcher Mimi Onuoha explains, "the grand ritual of collecting and reporting" data about the myriad forms of structural disadvantage Black people face in the United States has not resulted in meaningful change.[8] We should be careful not to over-ascribe power to data collection as enough, in itself, to effectuate reform. We also should recognize that these reform-oriented types of data reuse are, in a sense, only a silver lining to top-down corporate and government surveillance of workers—the part does not necessarily redeem the whole.

There is one other important role that technology may play here. Sometimes, the introduction of a technology into a social sphere can help snap problems into focus in a new way, providing an opportune moment to *reconsider* the actual order—and perhaps reconfigure it. My colleagues and I have described this phenomenon as "computing as synecdoche"[9]—the idea that technology can be a part that stands in for the whole, offering a new lens for making longstanding problems salient.

What if, for example, the introduction of the ELD caused us to reconsider everything about how trucking is organized and regulated, and called attention to its long-term struggles? To some extent, it has had this effect—the ELD has brought to the fore some aspects of trucking labor (particularly the prevalence of overwork) that were already the case, but which have often gone unnoticed or been taken for granted by the public. For instance, the strictness of electronic logging has certainly spurred discussion among regulators about the need to build more flexibility into the hours-of-service rules in the future—flexibility that has always been necessary, and that was previously achieved through logbook-fudging rather than meaningful reform to the rules themselves. This is a positive step forward that begins to account more realistically for the actual order of the trucking industry—but it is only a step. A more serious reckoning with the industry's root problems has yet to occur.

Technology isn't deterministic. The ways it affects work and workers depends on the details of how it's implemented, and alongside what other changes to the actual order of the social system. There are a number of hard questions we should ask. Are new tools introduced as a "shock" to the system, or gradually integrated with work practices over time?[10] What policies are put into place to govern the use of new tools, to allocate their risks, and to insulate workers from their harmful effects? Are workers retrained in new skills if their old skills become obsolete, and what social safety net—unemployment insurance, social support, health care provision—do we have in place to help them during periods of disruption? How can technologies be integrated into the identity and culture of an occupation, its workers' traditions, and goals *beyond* making money? How do we account for workers who are tasked with integrating their work *with* technological systems, not just being replaced by them? And perhaps the most important question of all: who gets to make these choices?

Most fundamentally, we have to recognize that successful technological change needs to be accompanied by social, economic, cultural, and legal reconfigurations if it has any hope of addressing the actual order. If we are to build tools that give managers oversight into workers' activities, we must accompany them with legal protections for those workers, and consider how they will (or won't) be integrated into the culture of the workplace.

A field experiment conducted by researchers at New York University illustrates the point.[11] The researchers found that truck drivers responded very differently to public posting of ELD-related performance metrics based on whether the workplace had a "collectivistic orientation"—that is, whether the workplace culture tended to emphasize teamwork and cooperation—as opposed to a more top-down culture that emphasized individual performance. In individualist workplace cultures, the researchers found that when everyone's performance was posted publicly, workers became more competitive with one another—ultimately driving performance up (for example, improving fuel efficiency). But interestingly, in a collectivist workplace, publicly posting performance metrics had precisely the *opposite* effect on worker performance, causing workers to perform *worse* on performance metrics than if each worker were only privy to their own data. The researchers conclude that this effect was a result of high performers in collectivist work cultures not wanting to cause low performers to feel bad, or to damage team spirit through competition. The point of the experiment is *not* to demonstrate that firms should avoid instilling collectivist values in the workplace if they want to spur their workers into competitive performance. Rather, the point is to recognize how different the effects of the monitoring technology, and the practices surrounding it, can be—depending on how people in the workplace relate to one another. Similarly, recall from chapter 6 how frequently ELD resistance practices function not necessarily to make truckers more money or to put them in a different material position, but instead to reassert their occupational autonomy and sense of self. Culture matters.

Economics also matter. Cracking down on rule-breaking without recognizing that workers are compelled to break rules in order to make a living doesn't change the economic structure that causes workers to break rules in the first place; it should be an occasion to consider broader reform. And similarly, building autonomous technologies that take away the only profitable parts of human workers' jobs can't happen in a vacuum; it must happen alongside reconsideration of *why* only certain parts of those jobs are profitable in the first place—and how it might be otherwise.

As digital technologies flourish across social life, they stand poised to reorient how people relate to institutions, to each other, and to themselves. Because technical, socioeconomic, and legal systems are so interdependent, a good deal of what we tend to think of as technological change doesn't directly involve technology at all. Often, the best way to think about technological change is not to focus solely on the technology, but to strengthen the social institutions and relationships that surround it.[12] Only by doing so can we ensure that digital technologies become part of a vibrant social order that protects workers and promotes human dignity.

Studying Surveillance

In this appendix, I provide some notes on the methodological approaches I took in my research. I describe the contours of multi-sited fieldwork in which I engaged, including details on the types of data sources I drew from in order to understand the ELD from multiple perspectives. I also explain some of the strategies I used in light of particular challenges that arose in my research. For instance, studying surveillance is uniquely complicated given that I, as a researcher, was myself acting as a kind of surveillor; once I realized this complexity in the field, I adjusted my research techniques accordingly (for instance, by asking questions in a more attenuated form to make participants more comfortable). I also offer some reflections on inter-actions between aspects of my personal identity and my communications with research participants, especially regarding gender dynamics.

It was crucial for me to approach the development, use, and meaning of electronic monitoring, as well as the trucking industry more generally, from the perspectives of multiple stakeholders. A multiple viewpoint approach is key to developing a nuanced understanding of how a technology functions in the world, why it does so, and who is benefitted or harmed by its use.[1] As we've discussed, the ELD is a multivalent technology, which means drasti-cally different things to different groups of people; it both impacts and is impacted by contentious economic, cultural, and legal logics. Therefore, in planning the research, I opted for an approach that would provide a breadth of viewpoints through distributed, multi-sited fieldwork.[2] I tried to select field sites and respondents that would allow me to understand the ELD in all its complexity.

This choice involved some trade-offs, as do all methodological deci-sions a researcher makes. For one, there is undoubtedly quite a bit about trucking that one can only learn from actually working in the industry as a trucker—which I did not do, but which scholars like Steve Viscelli and Anne Balay have.[3] I also could have opted for intensive single-sited fieldwork, or to spend time doing cross-country ride-alongs with a truck driver. Certainly those approaches would have taught me different things about trucking and ELDs—but I opted to forego that depth in favor of the breadth of multiple viewpoints.

Because truckers are constantly on the move (both physically, in the nomadic course of their work, and organizationally, between employment for different firms), it was not generally feasible for me to have sustained contact over time with individual drivers. Instead, my interactions with drivers tended to be singular "snapshot" encounters.[4] As such, I was not able to study longitudinally changes in particular drivers' career trajectories or engagement with the relevant technologies (for instance, talking with a particular driver both before and after an ELD was installed in his truck). Rather, I spoke with and observed many drivers at many times and locations to gather data about how the field as a whole was being affected by electronic monitoring.

In total, research for the study took me to eleven states, and took place in a wide range of settings: from truck cabs to law offices, corporate "war rooms" to roadside rest areas, fancy hotels to truck stop bars. My fieldwork, described below, was accompanied by seventy-seven unstructured interviews with respondents; these took place both face-to-face and by phone. These contacts were either obtained from fieldwork or based on referrals from prior participants. The bulk of the research was conducted between the summer of 2011 and the close of 2014, when the ELD mandate was being actively debated (but was not yet in effect); I continued to supplement my research through 2019.

Generally speaking, my data come from three interrelated spheres: truck drivers, trucking and technology firms, and legal and regulatory processes. I expand on my data sources in each sphere below.

Studying Truck Drivers

The experiences and viewpoints of truckers, as the targets of surveillance and regulation, are the primary focus of my research. I spoke with and observed truckers in a wide variety of places. I made multiple visits to large highway truck stops in two states, and spent four days among tens of thousands of truckers and trucking enthusiasts at the nation's largest trucking trade show. I interviewed drivers in break rooms at the firms where they worked, and chatted with them at additional, incidental sites, like trucking

supply stores. I typically approached drivers directly to ask them to participate in the research; in other cases, meetings were arranged with the help
of a third party, like the driver's firm or a personal contact.

I supplemented my firsthand observation with several other data sources
about drivers' experiences. I read hundreds of articles from trucker-oriented
media, including magazines, newsletters, and radio programs. I listened to,
and eventually appeared on, SiriusXM's Road Dog Trucking Radio (where I
was given the CB moniker "Road Scholar" by the hosts). I joined the Owner-
Operator Independent Drivers' Association and read the daily communications it sent to its members. I kept tabs on drivers' online activities, including YouTube channels and forums on which drivers shared information,
opinions, and complaints.

Studying Trucking and Technology Firms

I also examined ELDs from organizational perspectives, particularly those
of trucking companies and technology firms. During on-site field visits to
four trucking firms—including both top-ten carriers and very small operations—I talked to managers, safety directors, trip planners, dispatchers,
trainers, and other personnel. At the time, two of these firms used exclusively electronic logging systems; another was in the process of integrating
it into their operations; and another used exclusively paper-based systems.
Conducting my research during a period of transition, when ELDs were
becoming more common but weren't yet mandated, let me observe firsthand the social practices that attended analog vs. digital recordkeeping
mechanisms.

At one large firm, I sat in the company's central "war room" and was
permitted to listen in on dispatchers' real-time conversations with drivers.
I also met and spoke with trucking firm representatives at trade shows and
"listening sessions" held by the federal government, along with insurers, a
trucking historian, logistics consultants, and others whose work was related
to the industry in some way.

To understand the perspectives of technology firms that developed
ELDs, I attended fleet management conferences and seminars, a three-day
trucking technology summit, and several online training sessions presented
by ELD vendors for trucking companies. I spoke with trainers, account
managers, and consultants to learn about the capabilities and challenges
of deploying monitoring technology within trucking firms. I also reviewed

a wide range of relevant industry literature, including policy whitepapers, engineering reports, training materials, product manuals, and marketing literature dating back to the 1970s.

Studying Legal and Regulatory Processes

I draw as well from a variety of legal and regulatory sources, including analysis of thousands of documents related to four federal agency rulemakings involving ELDs, from 1987 through 2015. These documents include cost/benefit analyses conducted by federal agencies, studies on truck safety and driver fatigue, and public comments submitted by interested parties (drivers, trucking firms, representatives of interest groups, and many others) offering opinions and evidence about ELDs intended to influence pending regulatory decisions.[5] I obtained digital versions of many of these documents using regulations.gov, the online portal through which the public can submit comments on regulations; older documents were obtained manually from the National Archives in College Park, Maryland.

I also attended events where regulations were being debated: I observed three days of meetings of the Motor Carrier Safety Advisory Committee, an advisory board consisting of regulators and industry stakeholders, as well as two DOT-sponsored "listening sessions" (one in person and one via webcast) intended to solicit the opinions of drivers on pending regulations. I further considered judicial opinions, party pleadings, and audio recordings of oral arguments in ELD-related court cases, and interviewed lawyers and lobbyists working on ELD-related issues.

The law enforcement community was another area of focus, particularly for chapter 5, which focuses on the ELD's role in roadside truck inspections. I spent five days at an internal conference of the commercial vehicle enforcement community, where I spoke with dozens of troopers, sergeants, and representatives of state departments of transportation responsible for ELD-related enforcement. In addition, I observed truck inspections firsthand while embedded for two days in the commercial vehicle enforcement unit of a state police organization. I read law enforcement documents and viewed online seminars providing training for inspection processes. I spoke with dozens of truck drivers about their inspection experiences. I read regulatory documents in which members of the law enforcement community voiced policy opinions about the development of ELDs, and heard law enforcement representatives share their perspectives at regulatory advisory meetings and at federal listening sessions.

Research Strategies

Every research endeavor faces unforeseen challenges. In the field, qualitative researchers need to develop strategies (sometimes on the fly) to minimize biases in data collection and ensure that findings are rigorous, valid, and trustworthy.[6] In the interest of methodological transparency, and in the hopes that my own approaches might be useful to other researchers, I note here complexities that I encountered over the course of the research, how they affected the data I was able to collect, and the strategies I used to address them.

SURVEILLANCE AND STORYTELLING

Studying surveillance is a complicated proposition. Empirical studies of surveillance are often beset by the difficulty of observing a "zero state." Because the researcher, in asking questions of a research subject or monitoring behavior, is *themselves* acting as a surveillor of a sort, it can be complicated to determine what behavior under a wholly unmonitored state would look like—a methodological challenge akin to the Hawthorne effect, in which subjects modify their behavior (usually by improving it) as a reaction to the fact of being observed (often termed "observer effects").[7]

Observer effects can occur in any qualitative research context, but take on special significance in the context of studying surveillance. In my research, I quickly learned that asking participants many detailed questions about surveillance might make research participants uncomfortable and likely less forthcoming, particularly around issues of illegality and resistance. If a truck driver, for instance, were particularly sensitive to the idea that powerful parties were keeping track of his behaviors, my probing questions about his potential violation of legal rules—and perhaps even his illicit circumvention of technological controls—were sure to cause him to temper his responses in defense, and might also cause distress. Even though I was acting as an agent of neither the government nor any of the drivers' employers, I found that many drivers had a generalized distrust of authority and an unclear perception of what entities have access to what ELD-derived data.

These difficulties were somewhat exacerbated by my institutional affiliation when I began the research. I first interviewed drivers in the summer of 2011 when I was working as a research intern at Intel Corporation's Interaction and Experience Research Lab. To my knowledge, Intel had (and still has) no explicit business interest in ELDs; they permitted me to do the research under their auspices because of their interest in social aspects of technology use generally, whether those technologies were Intel-based or not. I explained this when I introduced myself to potential research participants and directed them to sign and initial a number of consent documents, as Intel required me to do. But I cannot help but think that it must have been a discordant and potentially intimidating experience when I explained that I was working for a technology company and wanted to ask sensitive questions about their use of a controversial and vaguely understood technology to circumvent the law. This particular difficulty resolved when my internship term ended, so that I no longer had to introduce myself with this affiliation and could revert to describing myself in terms of my academic position.

I used several tactics to allay any potential discomfort around my own position as a data collector. I affirmatively reminded participants about the confidentiality of our interactions, the fact that I wouldn't use their real names, and their ability to stop our conversation at any time. I initially anticipated audiotaping all of my interviews, but quickly noted that the presence of my recorder was a source of discomfort for surveillance-wary participants (not to mention that the audio quality was often unusably poor, since I tended to talk to respondents in truck stops and other noisy settings)—so I generally did away with audiotaping, instead taking detailed notes during and after interviews.

I also learned to rephrase questions in order to avoid putting participants on the spot about illegal or unsanctioned practices (for instance, instead of asking a driver "how do *you* cheat on your logbook?" I would ask "what types of things have you heard that *other drivers* do to falsify their logbooks?"). Allowing participants to attenuate their responses in this way enabled them to share information about their own practices without the risk of self-incrimination, and also had the benefit of providing me with a fuller picture of industry practices beyond those of the specific research participant.

Of course, this approach also potentially opened the door to less reliable data, based on hearsay rather than reported firsthand experience. Moreover, trucking as a profession has a rich tradition of storytelling, and trucker culture is rich with tales of roadside hijinks, clever workarounds in the face of technological or organizational ineptitude, and endurance of oppressive conditions. Some of these accounts certainly and intentionally deviate somewhat from reality, whether embellished for effect, strategically misstated in order to avoid incriminating oneself or others for unsanctioned behaviors, or for some other reason. In other cases, as I have noted, some research participants expressed (quite understandably) mistaken understandings about data flows within the ELD regime. Of course, the subjectivities and understandings of research participants are valuable in their own right, and these stories themselves thus provide an important and informative source of data distinct from other accounts.[8] Rumors about technologies embody a "process of collective interpretation"[9] about those artifacts, which demonstrates the capacities of user groups to reconfigure what technologies mean to them. Thus, truckers' accounts of (for instance) what ELDs could and could not track were important for me to understand the ELD as a cultural object. By no means do I intend to impugn or criticize my research participants in any way by noting these issues; rather, I raise them as a point of methodological import, and to explain how I thought about these types of accounts in analyzing my data.

Social researchers Torin Monahan and Jill Fisher contend that rather than thinking about observer effects as merely a form of "contamination" of a "supposedly pure social environment," we should recognize how these effects can be leveraged to generate new insights.[10] As they describe it, research participants attentive to the presence of a researcher may offer "staged performances," perhaps playing roles "over the top" for the perceived benefit of the audience (the researcher). But Monahan and Fisher point out that these performances are *themselves* a revealing source of data about how people *want to be perceived*. I think here of my exchange with a trucker recounted in chapter 4, in which he challenged me to name a destination in the United States so that he could demonstrate to me that he could, on the fly, describe the route without the benefit of GPS or a map. When he gave an accurate account and I asked how he'd learned the routes, he said proudly: "Honey, by driving them." This exchange demonstrated two things: first, that the trucker did, in fact, know these routes without reliance on external tools; but second (and perhaps even more importantly) that he *wanted to be perceived* by me as a capable, knowledgeable, skilled worker whose experience belied the need for technological oversight.

So these accounts were important sources of data in their own right, even if sometimes incorrect, exaggerated, or described differently on account of my presence. But in order to make my own evaluations about industry practices, I also needed to disentangle my research participants' accounts from actual behaviors and capabilities as much as possible.[11] To do so, I depended on two primary strategies to probe the accuracy of my data.[12] First, I privilege accounts that I heard multiple times from different actors, especially differently situated actors. For instance, most of the methods of resistance that I detail in chapter 6 were shared with me by multiple drivers, and in many cases were also independently reported to me by trucking firms and law enforcement officers. Second, whenever possible, I triangulated participants' direct reports with data derived from other contexts, especially contexts in which the data were initially produced for a non-research purpose. Common sources for such triangulation included online bulletin boards (on which drivers conversed anonymously with one another about ELDs and related practices), industry press, technical specifications for products (for example, tamperproofing efforts provide confirmatory evidence that tampering occurs), and legal and regulatory documents. The more corroborative evidence I could gather from diverse contexts, the more confidence I derived about the existence of a given practice.

INTERACTIONS AND IDENTITY

The dimensions of a researcher's personal identity necessarily shape their interactions with research participants, as well as what the researcher can learn in the field.[13] My status as a young white woman researcher in a working-class environment dominated by men impacted the study in important ways—including, almost certainly, some of which I am not aware. In the interest of reflexivity and transparency, I share a few words here about how my own characteristics affected my interactions with research participants, as well as the course of the research more broadly.

Most of the people I spoke to in the study, particularly truckers, were men—reflecting the makeup of the profession. (I did make a point to talk with women and nonbinary truckers as well whenever possible.) That group was racially mixed but primarily white, and ranged in age from "baby truckers" in their early twenties to "old hands" in their eighties. Because trucking is a profession in which men predominate (and a particularly "macho" one at that), it is often assumed that women do not know much about how trucking works. This stereotype certainly held true for me when I began the project; I had no working knowledge of truck driving in terms of either its broad legal and economic organization or the day-to-day mechanics of trucking labor. As the research progressed, of course, my knowledge grew. Over time, I developed a better sense for the structure of the industry and the pressures inherent in trucking work, deciphered trucker slang, and worked out the technical and regulatory infrastructures underlying the push for electronic monitoring. But my youth (at least at the beginning of the study), race, and

gender facilitated my "playing dumb," to some degree, about these intrica-
cies in my interactions with drivers.

Allowing research participants to explain anew concepts that I understood enabled me to understand how they themselves understood, and
wanted to portray, key issues—without imposing my own impressions
onto the interaction. This was essential for uncovering mistaken perceptions, such as those about the monitoring capabilities of electronic devices.
It also served a more instrumental purpose by easing my interactions and
building rapport. Many research participants, particularly truckers, had a
host of knowledge that they were very happy to share with whomever is
interested enough to listen; explaining the ins and outs of their work to a
newcomer seemed to be enjoyable for them. Respondents generally seemed
fairly candid and enthusiastic to share their experiences. Though I lack a
counterfactual, I would surmise that many of my research participants were
more likely to assent to spending time talking to a young white woman than
they would have been if I had had different characteristics, and I recognize
that my race, age, and gender greatly facilitated access in many ways.

I never lied to research participants about who I was or why I was
interested in them, and I am confident that I did not abuse the trust of any
research participant. But I was selective in what I revealed to them about
myself, and how. In addition to sometimes "playing dumb" about my existing
knowledge about the trucking industry, I only rarely mentioned my status
as a lawyer, which I think would have changed how participants viewed
me. (I felt it was acceptable to omit this description because while my legal
training informs my scholarship, I am not in active legal practice.)

My gender, in particular, sometimes served as a methodological obstacle
as well as an asset. Truckyards can be loud and disorienting places with
limited visibility, and I had to think about my personal safety in doing my
research. Some of the trucking firms and other places I visited were also
concerned about my safety (and presumably their potential liability), and
so would only let me talk to drivers or walk around truckyards if I brought
a male companion with me for security reasons. Though I was grateful for
the protection, this also limited my access: for instance, if my "guard" were
unavailable, I could not accompany drivers to their trucks to see their electronic equipment firsthand. (While this was not always necessary—the great
majority of my observations and interviews could be conducted in a truck
stop, a conference room, or elsewhere—there were occasions where being
in the cab and having the driver explain something to me was particularly

informative.) A final twist arose when I became pregnant over the course of the fieldwork. As my pregnancy became outwardly apparent, it too both helped and hindered my research efforts. In some cases, particularly with research participants who were women, it helped me to establish personal rapport. In others, it was limiting: for instance, a law enforcement agency refused to let me observe truck inspections while I was pregnant; it also made it rather more difficult for me to have a beer with a research participant in a truck stop bar, as I had sometimes done previously.

———

Qualitative empirical research is always messy. It demands that the researcher be continually willing to reshape their methodological approaches in response to situations as they unfold. In addition, a researcher must be attentive to the ways in which their own identity shapes the direction of the research and their interactions with subjects. I have done my best here to explain how I thought about these issues during the research process, why I made the decisions that I did, and how I think those decisions influenced my data collection and analysis. Despite these challenges, multi-sited qualitative research was uniquely suited to developing a nuanced view of the meanings of electronic monitoring in the trucking industry.

Notes on the Organization of the Trucking Industry

For the uninitiated, modern trucking can be a confusing industry to understand, marked by a multitude of terms and acronyms that connote different operational strategies and employment arrangements. Trucking firms can be generally categorized by the length of a company's hauls. *Long-haul* operations (also called "over-the-road," or OTR) are generally interstate and occur over long distances, while *short-haul* operations are more local in scope (for legal purposes, restricted to within 150 air-miles of the truck's home base). The timekeeping rules and tracking technologies that are the focus of this book apply specifically to long-haul operations, in which fatigue is of much greater concern.

In addition, trucking companies can be categorized as *truckload* (*TL*), *less-than-truckload* (*LTL*), or *parcel* carriers. In truckload carriage, a truck is filled with one customer's freight at point A, and the entire freight is transported to and unloaded at point B (sometimes called "point-to-point" carriage). LTL carriers, on the other hand, haul multiple loads of freight, for several customers, in a single truck to multiple destinations, picking up and dropping off various loads periodically along a route; freight loads are sometimes brought to one of the company's terminals by one driver and then transferred onto a different truck with a different driver, in order to achieve operational efficiency. Finally, parcel carriers, like FedEx and UPS drivers, carry large numbers of very small deliveries (primarily packages for businesses and consumers) along regular routes, typically in smaller delivery trucks. TL carriers face greater market competition than do LTL carriers; LTL and parcel carriage require more established and complex infrastructure (including a preexisting network of terminals and a critical mass of customers), while TL carriage has far lower barriers to entry. In TL trucking, economic competition and route irregularity contribute to the management and fatigue issues on which I focus here.

I do not focus here on issues in parcel trucking. Most parcel drivers are exempted from the federal fatigue rules because they drive locally and are able to come home each night. However, technological management systems pervade parcel trucking as well, as firms try to achieve operational efficiencies and maximize driver productivity; in some ways, parcel truck drivers are subject to even more granular management strategies.

Yet another axis of division is between for-hire and private fleet trucking. Trucking companies that are primarily engaged in the business of moving freight (companies like Werner, Schneider, and many thousands of small carriers) are considered *for-hire* carriers; that is, they are primarily engaged in the business of selling trucking services to customers. But nearly half the freight hauled in trucking in the United States is moved by *private fleets*, which operate in-house fleets within firms that are not primarily engaged in trucking (for example, Walmart's trucking operations). In a sort of hybrid arrangement, trucking companies can also have *dedicated* operations, in which they haul much of the freight for a particular (non-trucking) company and often work out of on-site offices, but remain a separate organization. Though driver safety, fatigue, and turnover are issues in private fleets, they are generally somewhat safer and more stable than for-hire carriers: their drivers more commonly drive dedicated (regular) routes, and the pace of work is more predictable and optimizable. This book focuses primarily on for-hire carriers, though I also spoke with drivers and other representatives of the private-fleet and dedicated segments.

Finally, the issues faced by employee drivers and owner-operators differ somewhat. Both are subject to the same legal rules relating to work hours. And both are under tremendous pressures to maximize their driving time, though the nature of those pressures differs. Truckers who work as employee drivers may face direct pressure from their employers, who operate on razor-thin margins in a highly competitive industry, to break the law by staying on the road too long. In trucking, on-time delivery to the customer is paramount; adherence to federal safety rules is frequently trumped by companies' desire to maximize their employees' productivity by moving goods faster. In addition, employee drivers are far more likely to contend with fleet management systems (discussed in chapter 4) through which their employers seek to increase operational efficiency and managerial oversight of their activities. Owner-operators less commonly face such pressures, but can lose out on opportunities for work altogether if they hew too closely to the rules. Particularly among owner-operators who negotiate each haul separately, there is strong economic pressure to accept a good haul offer; there is almost always someone else willing to break the rules to do so. My analysis in this book considers both employee and owner-operator truckers.

ACKNOWLEDGMENTS

First and foremost, I am immensely grateful to Meagan Levinson and the editorial staff at Princeton University Press for their support, advice, patience, and skill. I couldn't have asked for a better experience than working with Meagan and the rest of the PUP staff in bringing this project to fruition. Particular thanks to Nathan Carr, Jackie Delaney, and Michele Rosen for their guidance and hard work in turning my manuscript into a book.

The seeds of this research were planted at Intel Labs in Hillsboro, Oregon, where I spent the summer of 2011 as an intern. Intel's researchers were incredibly (and unexpectedly) supportive of my inkling that hanging out at truck stops would be a good use of my time. On an airplane from Portland to Santa Clara, I made Alex Zafiroglu listen to my recounting of an NPR story about trucking regulations; her reply that the story was "a little interesting" was all the encouragement I needed. From there, Dawn Nafus, Tad Hirsch, and Ken Anderson provided a perfect balance of guidance and freedom to let me run with the project, which subsequently became my dissertation, and then this book. It wouldn't exist without them.

My mentors at Princeton University's Department of Sociology—Paul DiMaggio, Matt Salganik, Kim Lane Scheppele, and Janet Vertesi—provided indispensable guidance, shaped my understanding of what it means to do rigorous, meaningful social scientific research, and taught me that asking the right questions is infinitely more valuable than answering the wrong ones. My postdoctoral mentors danah boyd and Helen Nissenbaum opened doors for me that I never expected to get to walk through. My law school mentor Kevin Emerson Collins was the first to spur my interest in thinking about the relationship between technology and rules, and Judge Sarah Evans Barker taught me to write about those rules when I clerked in her court. I wouldn't be where I am without the generosity, friendship, and advice of these incredible people, and I hope to be half the mentor to my own students that they have been to me.

My students and colleagues at Cornell have provided a wonderfully generative environment for completing this book. The Department of Information Science is a unique intellectual community that privileges multiple ways of knowing and has always made me feel valued both as a researcher and as a whole person. I'm particularly grateful to my students and to the members of the Artificial Intelligence, Policy, and Practice Initiative, as well as to Solon Barocas, James Grimmelmann, Steve Jackson, Jon Kleinberg, Wendy Ju, Mor Naaman, Helen Nissenbaum, Phoebe Sengers, and Malte Ziewitz, for many helpful conversations about this research.

I'm immensely lucky to be a member of a wonderfully supportive scholarly community which has improved my work immeasurably. I'm particularly thankful for generous feedback from Ifeoma Ajunwa, Mike Ananny, Anne Balay, Alvaro Bedoya, Michael Belzer, Yochai Benkler, Dan Bouk, danah boyd, Sarah Brayne, Kiel Brennan-Marquez, Ryan Calo, Angèle Christin, Danielle Citron, Julie Cohen, Kate Crawford, Madeleine Elish, Virginia Eubanks, Marion Fourcade, Tarleton Gillespie, Sue Glueck, Mary Gray, Woody Hartzog, Keith Hiatt, Chris Hoofnagle, Ian Kerr, Sarah Lageson, Kristian Lum, Mary Madden, Alexis Madrigal, Alice Marwick, Deirdre Mulligan, Paul Ohm, Frank Pasquale, Catherine Powell, David Robinson, Brishen Rogers, Matt Salganik, Eric Sears, Andrew Selbst, Evan Selinger, Scott Skinner-Thompson, Luke Stark, Kathy Strandburg, Daniel Susser, Steve Viscelli, and Ari Waldman.

At many stages of this project, writing groups were crucial to both my productivity and my sanity. I'm grateful to Naomi Adiv, Joel Correia, Liz Derickson, Anna Lauren Hoffmann, Meg Leta Jones, Nick Klein, Amy Krosch, Amanda Levendowski, Lindsey Novak, Rourke O'Brien, Marie Ostby, David Pedulla, and Dory Kornfeld Thrasher for being fellow travelers who gave thoughtful feedback and advice, and whose own brilliance lit the fire for me to put words on paper on many occasions. All of them have heard more about the trucking industry than they bargained for over the past several years; I'm grateful for their friendship and indulgence.

Jason Schultz graciously informed me that my original title for the book, *The Automation of Compliance*, was kind of a snooze, and suggested that I engage in a "data-driven" process to brainstorm a catchier title; thanks to his felicitous phrasing, I didn't need to! I'm greatly indebted to Pegah Moradi, Jessie Taft, and Amanda Kaplowitz, whose expertise and hard work as research assistants greatly enriched this book.

This book has benefitted from many iterations of workshopping and presentation. I'm thankful for generous feedback from my hosts and participants at events held at the Canadian Institute for Advanced Research (CIFAR), Carleton College, Carnegie Mellon University, Columbia University, Cornell University, Data & Society Research Institute, Fashion Institute of Technology, Fordham University, Georgetown University, Hunter College, Indiana University, Microsoft Research, the Montreal Speaker Series on AI Ethics, the New School for Social Research, New York University, North Carolina State University, Notre Dame University, Princeton University, Rochester Institute of Technology, Rutgers University, SXSW, Transportation Research Forum, University College London, University of California–Berkeley, University of California–Davis, University of Michigan, University of North Carolina–Charlotte, University of Pennsylvania, and Yale University. I also thank the participants at the Privacy Law Scholars Conference, the Northeast Privacy Scholars Workshop, the We Robot Conference, the American Sociological Association Annual Meeting, the Law and Society Association Annual Meeting, the Security and Human Behavior Workshop, and the Annual Meeting of the Special Interest Group for Computing, Information, and Society (SIGCIS).

Parts of this book are adapted from my doctoral dissertation at Princeton University, "The Automation of Compliance: Techno-Legal Regulation in the United States Trucking Industry," as well as previous publications. Portions of chapters 2 and 3 were published in "Digital Surveillance in the Hypermasculine Workplace," *Feminist Media Studies* 16, no. 2 (2016): 361–65. An early version of chapter 5 was published as "The Contexts of Control: Information, Power, and Truck-Driving Work," *The Information Society* 31, no. 2 (2015): 160–74.

I'm grateful for generous financial support from the National Science Foundation (award #1228436), the John D. and Catherine T. MacArthur Foundation, the Alfred P. Sloan Foundation, the Horowitz Foundation for Social Policy, a New America National Fellowship, the Center for the Future of Arizona, the Cornell Center for Social Sciences, Intel Labs, and Microsoft.

My family—especially Roberta Levy, Morris and Mechas Levy, Dinah and John Fuller, Mike and Gigi Conway, Susan Levy, and Cooper Smith—has been steadfastly supportive and helped me keep a sense of humor about myself and this project. My husband Sam has been totally unwavering in his support, enthusiasm, and patience; I couldn't have done this, or most things, without him. And my children Sadie and Benny make this book only the third most significant production of my life so far. I'm so thankful for their encouragement and love.

Finally, I am exceptionally grateful to the many members of the trucking community—drivers, dispatchers, state troopers, technologists, and everyone in between—whose experiences are the basis of this book, as well as the members of the trucking trade press who painstakingly document the issues affecting the industry and its workers. I was overwhelmed by how generously and patiently members of the trucking community shared their time and resources with me, welcoming me into their homes, trucks, and offices across the country, to little immediate benefit of their own. I can only hope that, by sharing their stories as fairly and as honestly as I can, I can begin to repay my great debt to them in some small measure.

NOTES

Chapter 1. Introduction

1. Robert K. Merton, foreword to *The Technological Society*, by Jacques Ellul (New York: Vintage, 1964), vi.

2. Tyson Fisher, "What If the Transformers Had ELDs?" *Land Line Magazine* 43, no. 5 (July 2018): 52.

3. Fisher, 52.

4. In reality, the government routinely waives safety regulations, including the hours-of-service rules, for truckers aiding in disaster relief and other exigent circumstances; the Transformers would likely have been able to obliterate the Decepticons after regulatory intervention. "Emergency Declaration Information," Federal Motor Carrier Safety Administration, August 24, 2021, https://www.fmcsa.dot.gov/emergency-declarations#INFO.

5. "Five Trucking Companies Back EOBR Legislation," *Commercial Carrier Journal*, September 29, 2010, https://www.ccjdigital.com/business/article/14918249/5-trucking-companies-back-eobr-legislation. The quote is from Donald Osterberg, then senior vice president of safety for Schneider International. "EOBR" refers to an *electronic on-board recorder*, and was the commonly used name for an electronic logging device in the early and mid-2000s. Today, the federal government and the industry most commonly refer to the devices as ELDs; I use "ELD" terminology throughout for consistency and clarity.

6. The practice of pretextual traffic stops is frequently traced to racial profiling and increased violence toward, incarceration of, and financial penalties against Black drivers. Alexes Harris, "Daunte Wright and the Grim Financial Incentive behind Traffic Stops," *Vox*, April 15, 2021, https://www.vox.com/first-person/22384104/daunte-wright-police-shooting-black-lives-matter-traffic-stops; Gabriel J. Chin and Charles J. Vernon, "Reasonable but Unconstitutional: Racial Profiling and the Radical Objectivity of *Whren v. United States*," *George Washington Law Review* 83, no. 3 (2015): 882–942.

7. J. David Goodman and Matt Flegenheimer, "De Blasio's Vow to End Traffic Deaths Meets Reality of New York Streets," *New York Times*, February 14, 2014, https://www.nytimes.com/2014/02/15/nyregion/vow-to-end-traffic-deaths-vs-reality-of-city-streets.html.

8. Brian Napier, "Working to Rule—A Breach of the Contract of Employment?" *Industrial Law Journal* 1, no. 3 (1972): 125–34. Working to rule is an example of a phenomenon that legal scholars Jessica Bulman-Pozen and David E. Pozen call "uncivil obedience"—"hyperbolic, literalistic, or otherwise unanticipated adherence to . . . formal rules" for purposes of resistance and dissent. Jessica Bulman-Pozen and David E. Pozen, "Uncivil Obedience," *Columbia Law Review* 115, no. 4 (2015): 809.

9. John W. Meyer and Brian Rowan, "Institutionalized Organizations: Formal Structure as Myth and Ceremony," *American Journal of Sociology* 83, no. 2 (1977): 340–63; Paul J. DiMaggio

and Walter W. Powell, "The Iron Cage Revisited: Institutional Isomorphism and Collective Rationality in Organizational Fields," *American Sociological Review* 48, no. 2 (1983): 147–60.

10. Alvin W. Gouldner, *Patterns of Industrial Bureaucracy* (New York: The Free Press, 1954).

11. Michel Anteby, *Moral Gray Zones: Side-Production, Identity, and Regulation in an Aeronautic Plant* (Princeton: Princeton University Press, 2008).

12. Anteby, for example, notes that tolerance of homer-making gave managers the opportunity to grant differential leeway to individual employees in ways that they could not have using other mechanisms, like wages (which were strictly determined by a collective bargaining agreement)—granting managers "venues for far more nuanced regulation than corporate, company-wide policies might ever allow" and granting management "significant additional control" over workers. Anteby, 130.

13. This phenomenon is well-recognized in "gap studies"—a school of thought in the sociology of law, most popular in the 1960s and 1970s, focused on illustrating inconsistencies between law on the books and law in action. This field, an intellectual descendant of the legal realist movement, focused on the multitude of social factors that make real social practices around legal rules and institutions diverge from statutory prescription. Sociolegal gap studies became less popular in subsequent years, in large part due to criticism of one of its assumptions—namely, that law has rational goals and objectives that can be articulated, interpreted, and measured against. For a review of gap studies and critiques of the perspective, see Jon B. Gould and Scott Barclay, "Mind the Gap: The Place of Gap Studies in Sociolegal Scholarship," *Annual Review of Law and Social Science* 8 (2012): 323–35.

14. Quentin Hardy, "How Urban Anonymity Disappears When All Data Is Tracked," *New York Times*, April 19, 2014, https://bits.blogs.nytimes.com/2014/04/19/how-urban-anonymity-disappears-when-all-data-is-tracked/.

15. Jonathan L. Zittrain, *The Future of the Internet and How to Stop It* (New Haven: Yale University Press, 2008); Christina M. Mulligan, "Perfect Enforcement of Law: When to Limit and When to Use Technology," *Richmond Journal of Law and Technology* 14, no. 4 (2008): 1–49; Michael L. Rich, "Should We Make Crime Impossible?" *Harvard Journal of Law and Public Policy* 36, no. 2 (2013): 795–848.

16. Tarleton Gillespie, *Wired Shut: Copyright and the Shape of Digital Culture* (Cambridge, MA: MIT Press, 2007). Gillespie decries restrictions like these because "there is no room left for ambiguously permissible behavior; every use is either allowed or refused, free or priced, at the moment it is attempted, with no exceptions." Gillespie, 61.

17. To the extent rule violation becomes more difficult, it may also impair socially productive moral deliberation about what the rules should be, as well as productive civil disobedience that can bring about social change. Ian Kerr, "Digital Locks and the Automation of Virtue," in *From "Radical Extremism" to "Balanced Copyright": Canadian Copyright and the Digital Agenda*, Michael Geist, ed. (Toronto: Irwin Law, 2010): 247–303; Woodrow Hartzog et al., "Insufficiently Automated Law Enforcement," *Michigan State Law Review* 2015, no. 5 (2015): 1763–96.

18. They accomplish this with questionable effectiveness. Bryce Newell, *Police Visibility: Privacy, Surveillance, and the False Promise of Body-Worn Cameras* (Oakland, CA: University of California Press, 2021).

19. Katherine C. Kellogg, Melissa A. Valentine, and Angèle Christin, "Algorithms at Work: The New Contested Terrain of Control," *Academy of Management Annals* 14, no. 1 (2020): 366–410.

20. For a thorough history of worker observation as a managerial tool, see Ethan S. Bernstein, "Making Transparency Transparent: The Evolution of Observation in Management Theory," *Academy of Management Annals* 11, no. 1 (2017): 217–66.

21. Bobby Allyn, "Your Boss Is Watching You: Work-From-Home Boom Leads to More Surveillance," *NPR*, May 13, 2020, https://www.npr.org/2020/05/13/854014403/your-boss-is-watching-you-work-from-home-boom-leads-to-more-surveillance.

22. Jodi Kantor, Karen Weise, and Grace Ashford, "The Amazon That Customers Don't See," *New York Times*, June 15, 2021, https://www.nytimes.com/interactive/2021/06/15/us/amazon-workers.html.

23. Jessica Vitak and Michael Zimmer, "Workers' Attitudes toward Increased Surveillance during and after the Covid-19 Pandemic," Social Science Research Council *Items* (blog), September 16, 2021, https://items.ssrc.org/covid-19-and-the-social-sciences/covid-19-fieldnotes/workers-attitudes-toward-increased-surveillance-during-and-after-the-covid-19-pandemic/; Jack Morse, "Microsoft Waters Down 'Productivity Score' Surveillance Tool After Backlash," *Mashable*, December 1, 2020, https://mashable.com/article/microsoft-365-productivity-score-workplace-surveillance-backlash.

24. Alex Rosenblat and Luke Stark, "Algorithmic Labor and Information Asymmetries: A Case Study of Uber's Drivers," *International Journal of Communication* 10 (2016): 3758–84.

25. Karen Levy and Solon Barocas, "Refractive Surveillance: Monitoring Customers to Manage Workers," *International Journal of Communication* 12 (2018): 1166–88.

26. Patricia Sánchez Abril, Avner Levin, and Alissa Del Riego, "Blurred Boundaries: Social Media Privacy and the Twenty-First-Century Employee," *American Business Law Journal* 49, no. 1 (2012): 63–124.

27. Ifeoma Ajunwa, Kate Crawford, and Jason Schultz, "Limitless Worker Surveillance," *California Law Review* 105, no. 3 (2017): 735–76.

28. Abubakar Bello Garba et al., "Review of the Information Security and Privacy Challenges in Bring Your Own Device (BYOD) Environments," *Journal of Information Privacy and Security* 11, no. 1 (2015): 38–54.

29. Shane Hamilton, *Trucking Country: The Road to America's Wal-Mart Economy* (Princeton: Princeton University Press, 2008).

30. R. W. Connell, *Masculinities* (Berkeley, CA: University of California Press, 1995), 55.

31. Robert Merton defined a strategic research site as one that "exhibits the phenomena to be explained or interpreted to such advantage and in such accessible form that it enables the fruitful investigation of previously stubborn problems and the discovery of new problems for further inquiry." Robert K. Merton, "Three Fragments from a Sociologist's Notebooks: Establishing the Phenomenon, Specified Ignorance, and Strategic Research Materials," *Annual Review of Sociology* 13, no. 1 (1987): 10–11.

32. Compare Tarleton Gillespie's characterization of the "heterogenous square" of technological control in the case of antipiracy copyright regimes as a mutually bolstering regulatory system of "technical artifact, commercial agreement, cultural justification, [and] legal authority." Gillespie, *Wired Shut*, 17. Gillespie emphasizes the need to analyze "regimes of alignment" surrounding technologies, which he describes as the "prominent inclusion of technical elements into an aligned set of efforts, partnerships, and laws that together arrange people and activities into a coherent system . . . [through] the interlocking of the technological, the legal, the institutional, and the discursive." Gillespie, 99, 102. Julie Cohen's work on the political economy of information capitalism also provides important theoretical context here. Cohen emphasizes that contemporary accounts of law and technology simply as opposing forces are insufficiently nuanced; rather, she demonstrates that the very institutions of law themselves are being altered by networked information technologies, notably with respect to the derivation of value from data about people and their activities (what Cohen terms the "biopolitical public domain"). Julie E. Cohen, *Between Truth and Power: The Legal Constructions of Information Capitalism* (Oxford: Oxford University Press, 2019).

33. Diane E. Bailey, Paul M. Leonardi, and Stephen R. Barley, "The Lure of the Virtual," *Organization Science* 23, no. 5 (2012): 1485–1504; Stephen R. Barley, "Technology as an Occasion for Structuring: Evidence from Observations of CT Scanners and the Social Order of Radiology Departments," *Administrative Science Quarterly* 31 (1986): 78–108.

Chapter 2. If the Wheel Ain't Turnin', You Ain't Earnin': Trucker Politics, Economics, and Culture

1. American Trucking Associations, "Economics and Industry Data," archived at https://web.archive.org/web/20210424163909/https://www.trucking.org/economics-and-industry-data.

2. American Trucking Associations, "Economics and Industry Data."

3. "Occupational Employment and Wages, May 2020: 53–3032 Heavy and Tractor-Trailer Truck Drivers," United States Bureau of Labor Statistics, archived at https://web.archive.org/web/20210713164113/https://www.bls.gov/oes/current/oes533032.htm.

4. Ginger Strand, *Killer on the Road: Violence and the American Interstate* (Austin, TX: University of Texas Press, 2014).

5. Trip Gabriel, "Alone on the Open Road: Truckers Feel Like 'Throwaway People,'" *New York Times*, May 22, 2017, https://www.nytimes.com/2017/05/22/us/trucking-jobs.html.

6. "Knights on the Highway," promotional film produced by the Jam Handy Organization, 1938, archived at https://archive.org/details/Knightso1938.

7. Diane S. Owen, *Deregulation in the Trucking Industry* (Washington, DC: Federal Trade Commission Bureau of Economics, 1988).

8. Cynthia Engel, "Competition Drives the Trucking Industry," *Monthly Labor Review* 121, no. 4 (1998): 34–41.

9. Michael Belzer, *Sweatshops on Wheels: Winners and Losers in Trucking Deregulation* (Oxford University Press, 2000), p. 8.

10. Belzer, 21.

11. James Jaillet, "Trucker Pay Has Plummeted in the Last 30 Years, Analyst Says," *Overdrive*, March 7, 2016, https://www.overdriveonline.com/trucker-pay-has-plummeted-in-the-last-30-years-analyst-stays/.

12. "Occupational Employment and Wages, May 2020," United States Bureau of Labor Statistics.

13. Belzer, *Sweatshops on Wheels*.

14. Sean Smith and Patrick Harris, "Truck Driver Job-Related Injuries in Overdrive," United States Department of Labor Blog, August 17, 2017, archived at https://web.archive.org/web/20170427031745/https://blog.dol.gov/2016/08/17/truck-driver-job-related-injuries-in-overdrive.

15. Matt Cole, "Record Number of Truckers Killed in Workplace Fatalities in 2017," *Overdrive*, December 23, 2018, https://www.overdriveonline.com/record-number-of-truckers-killed-in-workplace-fatalities-in-2017/.

16. "Civilian Occupations with High Fatal Work Injury Rates," United States Bureau of Labor Statistics, archived at https://web.archive.org/web/20210126011345/https://www.bls.gov/charts/census-of-fatal-occupational-injuries/civilian-occupations-with-high-fatal-work-injury-rates.htm.

17. W. Karl Sieber et al., "Obesity and Other Risk Factors: The National Survey of U.S. Long-Haul Truck Driver Health and Injury," *American Journal of Industrial Medicine* 57, no. 6 (2014): 615–26.

18. Anne Balay and Mona Shattell, "PTSD in the Driver's Seat," *Atlantic*, March 22, 2016, https://www.theatlantic.com/health/archive/2016/03/long-haul-trucking-and-mental-health/474840/; Mona Shattell et al., "Occupational Stressors and the Mental Health of Truckers," *Issues in Mental Health Nursing* 31, no. 9 (2010): 561–68.

19. Dan Nosowitz, "The Long White Line: The Mental and Physical Effects of Long-Haul Trucking," *Pacific Standard*, June 14, 2017, https://psmag.com/social-justice/the-long-white-line-the-mental-and-physical-effects-of-long-haul-trucking.

20. Stephen V. Burks et al., "Trucking 101: An Industry Primer," *Transportation Research Circular* No. E-C146 (December 2010), http://onlinepubs.trb.org/onlinepubs/circulars/ec146 .pdf; Mark Schremmer, "Large Fleets Reduce Driver Count; Turnover Rate Hits 96%," *Land Line*, December 23, 2019, https://landline.media/large-fleets-reduce-driver-count-turnover-rate -hits-96/.

21. Jennifer Cheeseman Day and Andrew W. Hait, "Number of Truckers at All-Time High," United States Census Bureau, June 6, 2019, https://www.census.gov/library/stories/2019/06 /america-keeps-on-trucking.html.

22. Jeffrey Short, "White Paper: Analysis of Truck Driver Age Demographics Across Two Decades," American Transportation Research Institute (report), December 2014, https://truckingresearch.org/wp-content/uploads/2014/12/Analysis-of-Truck-Driver-Age -Demographics-FINAL-12-2014.pdf.

23. Catie Edmondson, "'What Does a Trucker Look Like?' It's Changing, Amid a Big Short-age," *New York Times*, July 28, 2018, https://www.nytimes.com/2018/07/28/us/politics/trump -truck-driver-shortage.html.

24. American Trucking Associations, "Driver Shortage Update 2021," October 25, 2021, https://www.trucking.org/sites/default/files/2021-10/ATA%20Driver%20Shortage%20 Report%202021%20Executive%20Summary.FINAL_.pdf.

25. Steve Viscelli, *The Big Rig: Trucking and the Decline of the American Dream* (Oakland, CA: University of California Press, 2016), p. 8.

26. Edmondson, "'What Does a Trucker Look Like?'"

27. Mark Schremmer, "It's Time to Treat the Root Cause of Driver Retention Problem," *Land Line*, November 24, 2021, https://landline.media/its-time-to-treat-the-root-cause-of-driver -retention-problem/.

28. Cristina Roca and Dieter Holger, "Drawn by the Salary, Women Flock to Trucking," *Wall Street Journal*, October 14, 2019, https://www.wsj.com/articles/drawn-by-the-salary-women-flock -to-trucking-11571045406.

29. Anne Balay, *Semi Queer: Inside the World of Gay, Trans, and Black Truck Drivers* (Chapel Hill, NC: University of North Carolina Press, 2018).

30. Mary Pilon, "Surviving the Long Haul," *New Republic*, July 12, 2016, https://newrepublic .com/article/135000/surviving-long-haul.

31. Rip Watson, "Driver Demographics Begin to Shift," *Transport Topics*, December 11, 2014, https://www.ttnews.com/articles/driver-demographics-begin-shift.

32. Dale Belman, Francine Lafontaine, and Kristen A. Monaco, "Truck Drivers in the Age of Information: Transformation Without Gain," in *Trucking in the Age of Information*, eds. Dale Belman and Chelsea White III (London: Routledge, 2005), 183–212.

33. Quoctrung Bui, "Map: The Most Common Job in Every State," NPR.org, February 5, 2015, http://www.npr.org/sections/money/2015/02/05/382664837/map-the-most-common -job-in-every-state.

34. Viscelli, *Big Rig*, 9.

35. American Trucking Associations, "Reports, Trends & Statistics," 2019, archived at https:// web.archive.org/web/20200128011852/https://www.trucking.org/News_and_Information _Reports_Industry_Data.aspx.

36. Steve Viscelli, "Truck Stop: How One of America's Steadiest Jobs Turned into One of Its Most Grueling," *Atlantic*, May 10, 2016, https://www.theatlantic.com/business/archive/2016 /05/truck-stop/481926/.

37. Loretta Chao, "Trucking Makes a Comeback, but Small Operators Miss Out," *Wall Street Journal*, September 23, 2015, https://www.wsj.com/articles/trucking-makes-a-comeback-but -small-operators-miss-out-1443050680.

38. A minority of firms *do* compensate drivers for nondriving time, or are beginning to experiment with it, particularly in light of the perceived shortage of labor in the industry. Nondriving compensation can occur either systematically or on an *ad hoc* basis; for instance, at some of my observations in trucking firms, dispatchers had discretionary authority to authorize drivers for additional pay in particular circumstances (such as total mechanical breakdowns), or to authorize them for per diem expenses (like meals, a truck stop shower fee, or a motel room). Some firms experiment with other financial arrangements, like weekly pay guarantees or payment per stop. But again, these are not typical arrangements. Todd Dills, "Guaranteed Pay Making Waves in Driver Compensation," *Overdrive*, July 2, 2018, https://www.overdriveonline.com/guaranteed-pay-making-waves-in-driver-compensation/.

39. For a thorough exploration of collective bargaining in trucking, see Viscelli, *Big Rig*.

40. Belzer, *Sweatshops on Wheels*, 44.

41. Viscelli, *Big Rig*.

42. Michael Belzer, "Building Back Better in Commercial Road Transport: Markets Require Fair Labor Standards," presentation to Motor Carrier Safety Advisory Council of Federal Motor Carrier Safety Administration, July 19, 2021, https://www.fmcsa.dot.gov/sites/fmcsa.dot.gov/files/2021-07/MCSAC%20Building%20Back%20Better%20In%20Commercial%20Road%20Transport%20-%20Markets%20Require%20Fair%20Labor%20Standards%20-%20July%202021_rev2.pdf.

43. Blake E. Ashforth and Glen E. Kreiner, "'How Can You Do It?': Dirty Work and the Challenge of Constructing a Positive Identity," *Academy of Management Review* 24, no. 3 (1999): 413–34.

44. The connection between masculinity and machinery, particularly regarding vehicles and driving, has been drawn in a number of other contexts, including drag racing, young drivers, and risky traffic behaviors. See, for instance, Amy L. Best, *Fast Cars, Cool Rides: The Accelerating World of Youth and Their Cars* (New York: NYU Press, 2006); Ben Chappell, *Lowrider Space: Aesthetics and Politics of Mexican American Custom Cars* (Austin, TX: University of Texas Press, 2012); Daniel Miller, *Car Cultures* (New York: Routledge, 2001). Trucking represents a particularly acute link since it is also deeply connected with notions of occupational autonomy and economic provision.

45. Anne Balay and Mona Shattell, "Long-Haul Sweatshops," *New York Times*, March 9, 2016, https://www.nytimes.com/2016/03/09/opinion/long-haul-sweatshops.html.

46. Hamilton, *Trucking Country*, 109–10.

47. My categorization of truckers into "cowboys" and "family men" is congruent with Stratford and colleagues' study of HIV risk factors among long-haul truckers, in which he and his colleagues characterize their subjects as risk-taking "highway cowboys," more moderate "old hands," and "Christian truckers/old married men," who exhibit the least risky behaviors and have the most stable family relationships. Dale Stratford et al., "Highway Cowboys, Old Hands, and Christian Truckers: Risk Behavior for Human Immunodeficiency Virus Infection among Long-Haul Truckers in Florida," *Social Science & Medicine* 50, no. 5 (2000): 737–49. Consider also Hamilton's characterization of popular trucking culture in the 1970s viewing "working-class manhood as a constant negotiation between the poles of promiscuity and fidelity." Hamilton, *Trucking Country*, 199.

48. Hamilton, *Trucking Country*, 214.

49. Belzer, *Sweatshops on Wheels*.

50. 49 C.F.R. Parts 300–399.

51. Mark Schremmer, "OOIDA on Improving Retention: 'Put It on the Paycheck,'" *Land Line*, December 17, 2021, https://landline.media/ooida-on-improving-retention-put-it-on-the-paycheck/.

Chapter 3. Tired Truckers and the Rise of Electronic Surveillance

1. Belzer, *Sweatshops on Wheels*, 37–38.

2. Belzer.

3. National Academies of Science, Engineering, and Medicine, *Commercial Motor Vehicle Driver Fatigue, Long-Term Health, and Highway Safety: Research Needs* (Washington, DC: The National Academies Press, 2016), https://www.nap.edu/catalog/21921/commercial-motor-vehicle -driver-fatigue-long-term-health-and-highway-safety.

4. In 2019, 5,005 people were killed in truck crashes, and 159,000 were injured. These numbers include truckers and other truck occupants, drivers and occupants of other vehicles, and non-drivers (like pedestrians and cyclists). National Highway Traffic Safety Administration, "Traffic Safety Facts: 2019 Data," May 2021, https://crashstats.nhtsa.dot.gov/Api/Public/ViewPublication /813110.

5. Federal Motor Carrier Safety Administration Office of Research and Analysis, "The Large Truck Crash Causation Study—Analysis Brief," July 2007, https://www.fmcsa.dot.gov /safety/research-and-analysis/large-truck-crash-causation-study-analysis-brief; Tzuoo-Ding Lin, Paul P. Jovanis, and Chun-Zin Yang, "Modeling the Safety of Truck Driver Service Hours Using Time-Dependent Logistic Regression," *Transportation Research Record* iss. 1407 (1993): 1–10.

6. United States Department of Transportation, "U.S. Department of Transportation Declares Illinois Long-Haul Truck Driver to be an Imminent Hazard to Public Safety," February 12, 2014, https://www.transportation.gov/briefing-room/us-department-transportation-declares-illinois -long-haul-truck-driver-be-imminent-0.

7. Gail Sullivan, "Trucker in Tracy Morgan Crash Hadn't Slept for More Than 24 Hours," *Washington Post*, June 10, 2014, https://www.washingtonpost.com/news/morning-mix/wp/2014/06 /10/trucker-in-tracy-morgan-crash-hadnt-slept-for-more-than-24-hours/; Howard Abramson, "The Trucks Are Killing Us," *New York Times*, August 22, 2015, https://www.nytimes.com/2015 /08/22/opinion/the-trucks-are-killing-us.html.

8. Alan Derickson, *Dangerously Sleepy: Overworked Americans and the Cult of Manly Wakefulness* (Philadelphia: University of Pennsylvania Press, 2013).

9. Thomas J. Balkin et al., "The Challenges and Opportunities of Technological Approaches to Fatigue Management," *Accident Analysis & Prevention* 43, no. 2 (2011): 565–72.

10. Lawrence J. Ouellet, *Pedal to the Metal: The Work Lives of Truckers* (Philadelphia: Temple University Press, 1994), 134.

11. Ouellet, 136.

12. Derickson, *Dangerously Sleepy*, 114.

13. Derickson.

14. Derickson, 122.

15. Kathleen Frydl, *The Drug Wars in America, 1940–1973* (Cambridge: Cambridge University Press, 2013), 192–99.

16. Erica Goode, "Nerve Damage to Brain Linked to Heavy Use of Ecstasy Drug," *New York Times*, October 30, 1998, https://www.nytimes.com/1998/10/30/us/nerve-damage-to-brain -linked-to-heavy-use-of-ecstasy-drug.html.

17. "Police Investigate Truck Driver's Facebook Brag about 20 Hours of Driving," *CDL Life*, November 18, 2015, https://cdllife.com/2015/police-investigate-truck-drivers-facebook-brag -about-20-hours-of-driving/.

18. Brett Murphy, "Asleep at the Wheel," *USA Today*, December 28, 2017, https://www .usatoday.com/pages/interactives/news/rigged-asleep-at-the-wheel/.

19. "Trucking Company Tells Tired Truck Drivers: Don't Stop When You're Tired . . ." *Trucking Watchdog*, July 6, 2016, https://www.truckingwatchdog.com/2016/07/06/trucking-company -tells-drivers-dont-stop-when-youre-tired/.

20. In so doing, truckers recall the working-class boys described by Willis, whose cultural practices support and re-entrench the very structures that ensure the continued exploitation of their labor. Paul Willis, *Learning to Labor: How Working Class Kids Get Working Class Jobs* (New York: Columbia University Press, 1981).

21. Balay, *Semi Queer*, 92.

22. Hours-of-service rules are listed in the Code of Federal Regulations, 49 C.F.R. § 395.3 (2020).

23. The "graph-grid" form was required and standardized by the Federal Highway Administration in 1982. See 47 Federal Register 53389 (November 26, 1982). The graph grid was intended to ensure uniformity in recordkeeping, thus easing enforcement of the hours-of-service rules.

24. Drivers who spend some off-duty time in the sleeper berth of the truck have some additional flexibility regarding how they split up required breaks with driving time. See 49 C.F.R. § 395.1(g) for a full explanation.

25. The method of recording a driver's duty status on the graph grid, and requirements for the filing and retention of drivers' logs, are set forth at 49 C.F.R. §§ 395.8(h), (i), and (k).

26. Dale L. Belman and Kristen A. Monaco, "The Effects of Deregulation, De-Unionization, Technology, and Human Capital on the Work and Work Lives of Truck Drivers," *Industrial and Labor Relations Review* 54, no. 2A (2001): 502–24.

27. Elisa R. Braver et al., "Who Violates Work Hour Rules? A Survey of Tractor-Trailer Drivers," Insurance Institute for Highway Safety (report), January 1992, http://trid.trb.org/view .aspx?id=362329.

28. Owner-Operator Independent Drivers Association, Inc. v. Federal Motor Carrier Safety Administration, 656 F.3d 580 (7th Cir. 2011).

29. There is a fairly narrow set of exemptions to the ELD requirement for certain types of long-haul carriers. As I discuss in chapter 6, older trucks (model year 2000 and earlier) are exempt from the mandate because of concerns that these trucks would not accommodate ELD technology. Agricultural and livestock haulers are also exempt due to the need for additional flexibility for these carriers (for example, when hauling live animals).

30. "Survey Finds Larger Truck Fleets Ready for ELD Mandate; Small Firms, Not So Much," *DC Velocity*, September 14, 2016, https://www.dcvelocity.com/articles/28213-survey-finds-larger -truck-fleets-ready-for-eld-mandate-small-firms-not-so-much.

31. 80 Federal Register 78314 (December 16, 2015).

32. "Survey Finds Larger Truck Fleets Ready for ELD Mandate."

33. "Survey Finds Larger Truck Fleets Ready for ELD Mandate."

34. James Jaillet, "The New Cost of e-Log Compliance and FMCSA's Denial of Small Business Exemption," *Overdrive*, December 13, 2015, https://www.overdriveonline.com/the-new-cost-of -e-log-compliance-and-fmcsas-denial-of-small-business-exemption/.

35. Alan Smith and Forrest Lucas, "A New Electronic Logging Rule Could Drive Independent Truckers off the Road," *Hill*, October 17, 2017, https://thehill.com/opinion/energy-environment /355563-a-new-electronic-logging-rule-could-drive-independent-truckers-off.

36. Daren Hansen, "Trucking Alliance Calls for ELD Use by Intrastate Drivers," J. J. Keller Encompass (blog), July 26, 2018, https://eld.kellerencompass.com/resource/blog/trucking -alliance-calls-for-eld-use-by-intra-state-drivers.

37. Todd Dills, "The Not-So-Long Haul: The Operational Challenges to Owner-Ops after the ELD Mandate," *Overdrive*, July 22, 2021, https://www.overdriveonline.com/the-not-so-long-haul -the-operational-challenges-to-owner-ops-after-the-eld-mandate/.

38. Caroline Boris and Rebecca M. Brewster, "Managing Critical Truck Parking Case Study—Real World Insights from Truck Parking Diaries," American Transportation Research Institute (report), December 2016, https://truckingresearch.org/wp-content/uploads/2016/12/ATRI-Truck-Parking-Case-Study-Insights-12-2016.pdf, p. 15.

39. Boris and Brewster, 25.

40. Owner-Operator Independent Drivers Association, Inc. v. United States Department of Transportation, 840 F.3d 379 (7th Cir. 2016).

41. Ibid.

42. Todd Dills, "An ELD Positive, an Hours Negative, and More from Trucking's Present and Perhaps Stormy Future," *Overdrive*, July 29, 2018, https://www.overdriveonline.com/an-eld-positive-an-hours-negative-and-more-from-truckings-present-and-perhaps-stormy-future-in-expediting-panel/.

43. Pegah Moradi and Karen Levy, "The Future of Work in the Age of AI: Displacement or Risk-Shifting?" in *Oxford Handbook of Ethics of AI*, eds. Markus D. Dubber, Frank Pasquale, and Sunit Das (Oxford University Press, 2020), 271–88; Mary L. Gray and Siddharth Suri, *Ghost Work: How to Stop Silicon Valley from Building a New Global Underclass* (Boston: Houghton Mifflin Harcourt, 2019).

44. Karen Levy and Michael Franklin, "Driving Regulation: Using Topic Models to Examine Political Contention in the United States Trucking Industry," *Social Science Computer Review* 32, no. 2 (2014) :182–94.

45. "63% of Drivers are Detained More than 3 Hours Per Stop," *DAT Freight & Analytics*, July 12, 2016, https://www.dat.com/blog/post/63-of-drivers-are-detained-more-than-3-hours-per-stop.

46. Naomi J. Dunn et al., "Driver Detention Times in Commercial Motor Vehicle Operations," Federal Motor Carrier Safety Administration (report), December 2014, https://vtechworks.lib.vt.edu/bitstream/handle/10919/55062/13-060-Detention-508C-Dec14.pdf.

47. Erin Speltz and Dan Murray, "Driver Detention Impacts on Safety and Productivity," American Transportation Research Institute (report), September 2019, https://truckingresearch.org/wp-content/uploads/2019/09/ATRI-Detention-Impacts-09-2019.pdf, p. 11.

48. United States Department of Transportation Office of Inspector General, "Estimates Show Commercial Driver Detention Increases Crash Risks and Costs, but Current Data Limit Further Analysis," Report No. ST2018019, January 31, 2018, https://www.oig.dot.gov/sites/default/files/FMCSA%20Driver%20Detention%20Final%20Report.pdf.

49. Federal Motor Carrier Safety Administration, Hours of Service of Drivers: American Trucking Associations (ATA); Denial of Application for Exemption, 81 Federal Register 17240 (March 28, 2016).

50. Speltz and Murray, "Driver Detention Impacts on Safety and Productivity," 15.

51. Speltz and Murray.

52. Speltz and Murray, 14.

53. Speltz and Murray.

54. United States Department of Transportation Office of Inspector General, "Estimates Show Commercial Driver Detention Increases Crash Risks."

55. Ibid.

56. John Gallagher, "What Lawmakers Intend to Do about Driver Detention," *Freight Waves*, June 4, 2020, https://www.freightwaves.com/news/lawmakers-prod-fmcsa-to-move-on-driver-detention. FMCSA currently has enforcement authority over shippers and receivers if they coerce drivers into violating the hours-of-service regulations. Todd Dills, "How to Blow the Whistle on Problem Shippers/Receivers When Delays Force Violations," *Overdrive*, March 29, 2018, https://www.overdriveonline.com/channel-19/article/14894114/how-to-blow-the-whistle-on-problem-shippersreceivers-when-delays-force-violations.

57. Alex Scott, Andrew Balthrop, and Jason W. Miller, "Unintended Responses to IT-Enabled Monitoring: The Case of the Electronic Logging Device Mandate," *Journal of Operations Management* 67, no. 2 (2021): 152–81.

58. Todd Dills, "Stoking the Log Fires: Hours Violations, Fleet Size and ELDs," *Overdrive*, July 6, 2017, https://www.overdriveonline.com/stoking-the-log-fires/.

59. Scott, Balthrop, and Miller, "Unintended Responses to IT-Enabled Monitoring."

60. Tyson Fisher, "Truck Fatalities Highest in 30 Years in First Full Year of ELD Mandate," *Land Line*, October 23, 2019, https://landline.media/truck-fatalities-highest-in-30-years-in-first-full-year-of-eld-mandate/.

61. Steve Viscelli, "A Pragmatic View of How ELDs Are Changing Trucking," *Fleet Owner*, December 26, 2018, https://www.fleetowner.com/technology/telematics/article/21703331/a-pragmatic-view-of-how-elds-are-changing-trucking.

62. Mark Schremmer, "ELD Mandate Hasn't Reduced Crashes, Study Reveals," *Land Line*, February 5, 2019, https://landline.media/eld-mandate-hasnt-reduced-crashes-study-reveals/.

63. Scott, Balthrop, and Miller, "Unintended Responses to IT-Enabled Monitoring."

64. James Jaillet, "Study: Under ELDs, Crash Rates Flat but Unsafe Driving Violations Are Up," *Overdrive*, February 4, 2019, https://www.overdriveonline.com/electronic-logging-devices/article/14895792/study-under-elds-crash-rates-flat-but-unsafe-driving-violations-are-up.

65. Mark Schremmer, "Be a Minute Late, and There Will Be Hell to Pay," *Land Line*, May 9, 2018, https://landline.media/be-a-minute-late-and-there-will-be-hell-to-pay/.

Chapter 4. The Business of Trucker Surveillance

1. Kirstie Ball, "Workplace Surveillance: An Overview," *Labor History* 51, no. 1 (2010): 87–106; James R. Beninger, *The Control Revolution: Technological and Economic Origins of the Information Society* (Cambridge, MA: Harvard University Press, 1986).

2. Moradi and Levy, "The Future of Work in the Age of AI"; Kellogg, Valentine, and Christin, "Algorithms at Work."

3. The technology scholar Shoshana Zuboff terms this an "action-centered" approach to work. Shoshana Zuboff, *In the Age of the Smart Machine: The Future of Work and Power* (New York: Basic Books, 1988).

4. Phaedra Daipha, *Masters of Uncertainty: Weather Forecasters and the Quest for Ground Truth* (Chicago: University of Chicago Press, 2015).

5. Bailey, Leonardi, and Barley, "Lure of the Virtual."

6. Zuboff, *Age of the Smart Machine*; Harry Braverman, *Labor and Monopoly Capital: The Degradation of Work in the Twentieth Century* (New York: Monthly Review Press, 1974); Graham Sewell, "The Discipline of Teams: The Control of Team-Based Industrial Work through Electronic and Peer Surveillance," *Administrative Science Quarterly* 43, no. 2 (1998): 397–428; Jannis Kallinikos, *The Consequences of Information: Institutional Implications of Technological Change* (Cheltenham, UK: Edward Elgar Publishing, 2007); Cohen, *Between Truth and Power*.

7. Overdrive Staff, "PrePass Expands Bypass Capabilities, Introduces ELD," *Overdrive*, October 23, 2017, https://www.overdriveonline.com/prepass-expands-bypass-capabilities-introduces-eld/.

8. Matthew C. Camden et al., "Leveraging Large-Truck Technology and Engineering to Realize Safety Gains: Video-Based Onboard Safety Monitoring Systems," *AAA Foundation for Traffic Safety*, September 2017, https://aaafoundation.org/wp-content/uploads/2017/11/Truck-Safety_Video-Systems.pdf.

9. Omnitracs, *Omnitracs ELD Driver Retention Model*, https://www.omnitracs.com/sites/default/files/files/2018-10/ELD_Driver_Retention_Model_08_16_Brochure_web.pdf.

10. Federal Motor Carrier Safety Administration, *Regulatory Impact Analysis of Electronic On-Board Recorders*, November 2006, FMCSA-2004-18940-0352, p. 10.

11. Scott, Balthrop, and Miller, "Unintended Responses to IT-Enabled Monitoring."

12. Stratford et al., "Highway Cowboys, Old Hands, and Christian Truckers."

13. David Tanner, "'Why Aren't You Rolling?'" *Land Line Magazine*, December 2013: 32–33.

14. Jami Jones, "Special Report: OOIDA Brief Details Driver Harassment by ATA Member Companies," *Land Line*, March 6, 2012, archived at https://web.archive.org/web/20120307130122/http://www.landlinemag.com/Story.aspx?StoryID=23321.

15. Knut H. Rolland and Eric Monteiro, "Balancing the Local and the Global in Infrastructural Information Systems," *Information Society* 18, no. 2 (2002): 87–100; Bailey, Leonardi, and Barley, "Lure of the Virtual"; Katrin Jonsson, Jonny Holmström, and Kalle Lyytinen, "Turn to the Material: Remote Diagnostics Systems and New Forms of Boundary-Spanning," *Information and Organization* 19, no. 4 (2009): 233–52.

16. Barley, "Technology as an Occasion for Structuring"; Wanda J. Orlikowski, "The Duality of Technology: Rethinking the Concept of Technology in Organizations," *Organization Science* 3, no. 3 (1992): 398–427; Steven Blader, Claudine Gartenberg, and Andrea Prat, "The Contingent Effect of Management Practices," *Review of Economic Studies* 87, no. 2 (2020): 721–49.

17. John E. Mello and C. Shane Hunt, "Developing a Theoretical Framework for Research into Driver Control Practices in the Trucking Industry," *Transportation Journal* 48, no. 4 (2009): 20–39.

18. Chris Wolski, "Telematics Gamification Emphasizes Fun over 'Big Brother,'" *Automotive Fleet*, October 21, 2015, https://www.automotive-fleet.com/156349/telematics-gamification-emphasizes-fun-over-big-brother.

19. Michael Burawoy, *Manufacturing Consent: Changes in the Labor Process under Monopoly Capitalism* (Chicago: University of Chicago Press, 1979).

20. Overdrive Staff, "A Gold Rush for ELD Data," *Overdrive*, July 21, 2021, https://www.overdriveonline.com/electronic-logging-devices/article/14896446/a-gold-rush-for-eld-data.

21. For a comparison of major ELD vendors' data sharing policies, see Overdrive Staff, "Data Sharing Practices among Leading Providers of ELDs to Owner-Operators," *Overdrive*, June 17, 2019, https://www.overdriveonline.com/electronic-logging-devices/article/14896412/data-sharing-practices-among-leading-providers-of-elds-to-owner-operators.

22. Overdrive Staff, "ELD Data Handling: 'Privacy Is Paramount,' but Practices Vary," *Overdrive*, June 17, 2019, https://www.overdriveonline.com/electronic-logging-devices/article/14896447/eld-data-handling-privacy-is-paramount-but-practices-vary.

23. Leslie Scism, "America's Truckers Embrace Big Brother after Costing Insurers Millions," *Wall Street Journal*, June 4, 2017, https://www.wsj.com/articles/americas-truckers-embrace-big-brother-after-costing-insurers-millions-1496577601.

24. Overdrive Staff, "Gold Rush for ELD Data."

25. Mindy Long, "Truck Stops Step Up to Combat Parking Crisis," *Fleet Owner*, December 31, 2019, https://www.fleetowner.com/safety/article/21119313/truck-stops-step-up-to-combat-parking-crisis.

26. Todd Dills, "The Extent of Expanded Paid Parking Reservations after the ELD Mandate," *Overdrive*, July 22, 2021, https://www.overdriveonline.com/electronic-logging-devices/article/14894460/paid-parking-reservation-expansion-since-eld-mandate.

27. Todd Dills, "ELDs Up the Ante on Parking," *Overdrive*, June 2, 2018, https://www.overdriveonline.com/electronic-logging-devices/article/14894459/how-parking-has-changed-in-light-of-the-eld-mandate.

28. Overdrive Staff, "More Hopes and Hazards: ELD Data Aggregation for Advocacy, Predictive Maintenance, Other Uses," *Overdrive*, February 20, 2020, https://www.overdriveonline.com

/electronic-logging-devices/article/14896514/more-hopes-and-hazards-eld-data-aggregation-for
-advocacy-predictive-maintenance-other-uses.

29. Overdrive Staff, "More Hopes and Hazards."

30. Levy and Barocas, "Refractive Surveillance."

31. Todd Dills, "KeepTruckin to Leverage ELD Data for Freight Matching, Detention Pressure," *Overdrive*, April 17, 2019, https://www.overdriveonline.com/keeptruckin-to-leverage-eld
-data-for-freight-matching-detention-pressure/.

32. "The Extended Detention Exception," *KeepTruckin*, https://go.keeptruckin.com
/extended-detention-exception; Todd Dills, "ELD Maker Launches Petition to Add 14-Hour
Flexibility, Presents Data on Detention and Race-the-Clock Dynamics," *Overdrive*, November 19,
2017, https://www.overdriveonline.com/electronic-logging-devices/article/14893416/eld-maker
-launches-petition-to-add-14-hour-flexibility-presents-data-on-detention-and-race-the-clock
-dynamics.

33. Interoperability also underlies Haggerty and Ericson's notion of *the surveillant
assemblage*—the convergence of private and public, state and non-state component systems into
integrated, overlapping, continuous regimes of surveillance. Kevin D. Haggerty and Richard V.
Ericson, "The Surveillant Assemblage," *British Journal of Sociology* 51, no. 4 (2000): 605–22.

34. Federal Motor Carrier Safety Administration, *Proposed Rule: Electronic Logging Devices
and Hours of Service Supporting Documents*, March 28, 2014, FMCSA-2010-0167-0485.

Chapter 5. Computers in the Coop

1. Sarah Brayne, *Predict and Surveil: Data, Discretion, and the Future of Policing* (Oxford:
Oxford University Press, 2021); Newell, *Police Visibility*; Lauren Kilgour, "Designed to Shame:
Electronic Ankle Monitors and the Politics of Carceral Technology" (PhD dissertation, Cornell
University, 2021). Brayne, Newell, and Kilgour investigate the impacts of predictive analytics,
body-worn cameras, and electronic ankle monitors, respectively, with emphasis on how law
enforcement personnel interface with these technologies.

2. Brayne, *Predict and Surveil*, 75.

3. Overdrive Staff, "Hours of Service Violation Rate Cut in Half under ELD Mandate, FMCSA
Says," *Overdrive*, June 25, 2018, https://www.overdriveonline.com/electronic-logging-devices
/article/14894600/hours-of-service-violation-rate-cut-in-half-under-eld-mandate-fmcsa-says.

4. Bailey, Leonardi, and Barley, "The Lure of the Virtual"; Barley, "Technology as an Occasion for Structuring."

5. Some inspectors now do use "eRODS" (electronic record of duty status) software that
interfaces with ELD software to automatically flag violations—but even then, a human inspector
has discretion to decide whether to issue a citation.

6. "Things Your Drivers Don't Know About Roadside Inspections," *Oil Prophets: A Quarterly
Publication by the Petroleum & Convenience Marketers of Alabama* (Summer 2018): 18–19, https://
emflipbooks.com/flipbooks/PCMA/OilProhets_Magazine/Summer18/book/18/.

7. Todd Dills, "ELDs: What to Expect from Enforcement as the Mandate Descends,"
Overdrive, December 15, 2017, https://www.overdriveonline.com/elds-what-to-expect-from
-enforcement-as-the-mandate-descends/.

8. Todd Dills, "FMCSA Director Discusses ELDs and Enforcement: Know What You're Dealing with at Roadside," *Overdrive*, August 24, 2018, https://www.overdriveonline.com/fmcsa-on
-elds-and-enforcement-know-what-youre-dealing-with-at-roadside/.

9. Todd Dills, "E-Log Sticker Merits Easy Time at Weigh Station," *Overdrive*, February 12,
2017, https://www.overdriveonline.com/e-log-sticker-merits-easy-time-at-weigh-station/.

10. 49 C.F.R. §395.16(l)(1).

11. In some cases, physical separation could occur. Some ELD technologies allow data to be emailed or printed directly from the module. Wireless data transmission protocols are also increasingly used, obviating the need for an inspector to physically get into the truck. In other cases, the carrier's home office—which typically has access to the driver's logs in real time—may send data, upon request, to the police officer for inspection. Resort to these data transfer methods—some of which are decidedly low-tech, like faxing logs from the home office to the inspection station—are dependent on a number of factors: not all officers accept such transmissions as valid (according to the letter of the regulation, the driver must be able to produce the logs); an officer's own technological systems must be operating well (which they are often not); and officers who accept these logs must still confirm that the log that has been transmitted is, in fact, the driver's actual time record. The inseparability of ELD from truck has changed somewhat since my fieldwork was completed: subsequent ELD regulations allow ELDs to be based on a smartphone app that wirelessly connects to the vehicle's engine control module and that could be passed down out of the cab.

12. Todd Dills, "So Much for the 'Hope for Leniency': One Owner-Operator's Hours Out-of-Service Experience in the ELD Era," *Overdrive*, May 13, 2018, https://www.overdriveonline.com/so-much-for-the-hope-for-leniency-one-owner-operators-hours-out-of-service-experience-in-the-eld-era/.

13. Erving Goffman, *The Presentation of Self in Everyday Life* (New York: Anchor Books, 1959); Erving Goffman, "On Face-Work," in *Interaction Ritual: Essays on Face-to-Face Behavior* (Garden City, NY: Pantheon Books, 1967), 5–46.

14. Indiana State Police—Commercial Vehicle Enforcement Division, Facebook post, November 10, 2017, https://www.facebook.com/pg/ISPCVED/posts/.

15. Dills, "Stoking the Log Fires, Redux."

Chapter 6. Beating the Box: How Truckers Resist Being Monitored

1. Michel Foucault, *Discipline & Punish: The Birth of the Prison* (New York: Vintage Books, 1977).

2. Gary T. Marx, "A Tack in the Shoe: Neutralizing and Resisting the New Surveillance," *Journal of Social Issues* 59, no. 2 (2003): 369–90.

3. Finn Brunton and Helen Nissenbaum, "Vernacular Resistance to Data Collection and Analysis: A Political Theory of Obfuscation," *First Monday* 16, no. 5 (2011).

4. Steve Mann, Jason Nolan, and Barry Wellman, "Sousveillance: Inventing and Using Wearable Computing Devices for Data Collection in Surveillance Environments," *Surveillance and Society* 1, no. 3 (2003): 331–55.

5. Torin Monahan, "Counter-Surveillance as Political Intervention?" *Social Semiotics* 16, no. 4 (2006): 515–34.

6. James C. Scott, *Weapons of the Weak: Everyday Forms of Peasant Resistance* (New Haven: Yale University Press, 1987).

7. Helen Wells and David Wills, "Individualism and Identity Resistance to Speed Cameras in the UK," *Surveillance and Society* 6, no. 3 (2009): 259–74.

8. "HOW TO Hack or Cheat ANY ELOG in 60 Secs 'A MUST WATCH,'" YouTube video, December 16, 2017, https://www.youtube.com/watch?v=OM_rdM-Ym_k.

9. Radley Balko, "80 Percent of Chicago PD Dash-Cam Videos Are Missing Audio Due to 'Officer Error' or 'Intentional Destruction,'" *Washington Post*, January 29, 2016, https://www.washingtonpost.com/news/the-watch/wp/2016/01/29/80-percent-of-chicago-pd-dash-cam-videos-are-missing-audio-due-to-officer-error-or-intentional-destruction/.

10. Brayne, *Predict and Surveil*, 82.

11. Ross Anderson, "On the Security of Digital Tachographs," in *Proceedings of ESORICS 98: 5th European Symposium on Research in Computer Security* (1998), 111–25.

12. Rob Cave, "Rise in Lorry Tachograph Tampering on UK Roads," *BBC News*, September 24, 2017, http://www.bbc.com/news/uk-41361351.

13. "GPS Jamming: Out of Sight," *Economist*, July 27, 2013, https://www.economist.com /international/2013/07/27/out-of-sight.

14. Iain Thompson, "Feds Arrest Rogue Trucker after GPS Jamming Borks New Jersey Airport Test," *Register*, August 12, 2013, https://www.theregister.com/2013/08/12/feds_arrest_rogue _trucker_after_gps_jamming_disrupts_newark_airport/.

15. A. Ferreira et al., "How to Break Access Control in a Controlled Manner," in *Proceedings of the 19th IEEE International Symposium on Computer-Based Medical Systems* (2006): 847–54; Arvind Narayanan, "What Happened to the Crypto Dream? Part 2," *IEEE Security & Privacy* 11, no. 3 (2013): 0068–0071.

16. "How to Hack an MCP200 Truck PC and Play Games," YouTube video, March 2, 2012.

17. Patricia Ewick and Susan Silbey, "Narrating Social Structure: Stories of Resistance to Legal Authority," *American Journal of Sociology* 108, no. 6 (2003):1331.

18. Diane Vaughan, *Controlling Unlawful Organizational Behavior: Social Structure and Corporate Misconduct* (Chicago: University of Chicago Press, 1985), 60 (quoting Arthur L. Stinchcombe, "Social Structure and Organizations," in *Handbook of Organizations*, ed. James March [Routledge, 1975]).

19. Gary E. Marchant, "The Pacing Problem," in *The Growing Gap between Emerging Technologies and Legal-Ethical Oversight*, eds. Gary E. Marchant, Braden R. Allenby, and Joseph R. Herkert (New York: Springer, 2011), 199–205.

20. The term "grandfather clause" has racist origins. The original grandfather clauses were exemptions from stringent Jim-Crow-era laws enacted to prevent former slaves and their descendants from voting, like the provision of literacy tests and poll taxes. Grandfather clauses exempted from these requirements those whose ancestors had previously had the right to vote—in other words, white Southerners and their descendants. In this way, grandfather clauses had the effect of applying these restrictive voting hurdles only to Black people. Alan Greenblatt, "The Racial History of the 'Grandfather Clause,'" *NPR*, October 22, 2013, https://www.npr.org/sections /codeswitch/2013/10/21/239081586/the-racial-history-of-the-grandfather-clause.

21. Todd Dills, "Eyes on the Prize: The Market for Pre-2000 Trucks in Light of ELD Exemption," *Overdrive*, December 15, 2016, https://www.overdriveonline.com/eyes-on-the-prize-the -market-for-pre-2000-trucks-in-light-of-eld-exemption/.

22. Todd Dills, "Older Trucks Holding Value Better since ELD Announcement," *Overdrive*, December 9, 2016, https://www.overdriveonline.com/older-trucks-holding-value-better-since -eld-announcement/.

23. Eric Lipton, "How $225,000 Can Help Secure a Pollution Loophole at Trump's E.P.A.," *New York Times*, February 15, 2018, https://www.nytimes.com/2018/02/15/us/politics/epa -pollution-loophole-glider-trucks.html.

24. Eric Lipton, "'Super Polluting' Trucks Receive Loophole on Pruitt's Last Day," *New York Times*, July 6, 2018, https://www.nytimes.com/2018/07/06/us/glider-trucks-loophole-pruitt.html.

25. "eFleetSuite—FMCSA 395 Subpart B Compliant Electronic Driver Logs Application." Trimble IA Fleet Services, https://store.isefleetservices.com/Articles.asp?ID=259.

26. Todd Dills, "One20 Launches the F-ELD," *Overdrive*, March 24, 2017, https://www .overdriveonline.com/one20-launches-the-f-eld/.

27. Todd Dills, "Finger on the ELD Mandate—Good Marketing or Well over the Line?" *Overdrive*, September 17, 2017, https://www.overdriveonline.com/finger-on-the-eld-mandate-good -marketing-or-well-over-the-line/.

28. Albert O. Hirschman, *Exit, Voice, and Loyalty: Responses to Decline in Firms, Organizations, and States* (Cambridge, MA: Harvard University Press, 1970).

29. Howard S. Becker, "Notes on the Concept of Commitment," *American Journal of Sociology* 66 (1960): 3240.

30. Todd Dills, "ELD Mandate: Independents' Final Straw?" *Overdrive*, April 17, 2014, https://www.overdriveonline.com/electronic-logging-devices/article/14886211/eld-mandate-independents-final-straw.

31. Todd Dills, "ELD Headaches: Dealing with Technical Glitches and Other Equipment Issues," *Overdrive*, March 27, 2018, https://www.overdriveonline.com/eld-headaches-dealing-with-technical-glitches-and-other-equipment-issues/.

32. Overdrive Staff, "What ELD Exodus? Major Broker's Anecdotal Take on Capacity Constraints under Mandate," *Overdrive*, August 8, 2018, https://www.overdriveonline.com/what-eld-exodus-major-brokers-anecdotal-take-on-capacity-constraints-under-mandate/.

33. United States Department of Transportation, "Analysis of Driver Critical Reason and Years of Driving Experience in Large Truck Crashes," January 2017, https://rosap.ntl.bts.gov/view/dot/31693.

34. Associated Press, "Most Truckers End 11-Day Strike," *New York Times*, February 12, 1974, https://www.nytimes.com/1974/02/12/archives/most-truckers-end-11day-strike-traffic-called-nearly-normal-despite.html.

35. FOX59 Web Staff, "'Slow Roll' Rolls through Indy without Incident as Truckers Protest Regulations," *FOX59*, February 21, 2019, https://fox59.com/2019/02/21/police-warn-truckers-ahead-of-slow-roll-protest-around-i-465-thursday/.

36. James Jaillet, "Protestors Reckon with Minimal 'Shutdown' and Protest Participation," *Overdrive*, April 20, 2019, https://www.overdriveonline.com/after-april-12-protests-fail-to-materialize-organizers-plan-smaller-relaunch-of-facebook-efforts/.

37. 49 C.F.R. § 395.11 (2017).

38. Aaron K. Martin, Rosamunde E. Van Brakel, and Daniel J. Bernhard, "Understanding Resistance to Digital Surveillance: Towards a Multi-Disciplinary, Multi-Actor Framework," *Surveillance and Society* 6, no. 3 (2009): 213–32.

39. Gillespie, *Wired Shut*.

40. Todd Dills, "Poll: Since the ELD Mandate, Have You Felt Pressure from Carriers/Customers to Violate Hours Regs More or Less Often?" *Overdrive*, March 26, 2019, https://www.overdriveonline.com/poll-since-the-eld-mandate-have-you-felt-pressure-from-carriers-customers-to-violate-hours-regs-more-or-less-often/.

41. Journalists have documented similar dynamics involving Amazon delivery firms. Amazon requires its delivery drivers to use an app called Mentor that monitors drivers' behaviors and gives them safety scores, akin to the fleet management systems discussed in chapter 4. In 2021, *Vice* reported that these drivers were being told by their employers—companies that contract with Amazon to deliver packages, rather than Amazon itself—that they should turn Mentor on for a few hours at the beginning of the shift and then turn it off. Presumably, firms gave drivers these instructions so that they could both meet Amazon's requirements *and* encourage drivers to drive recklessly enough to meet delivery quotas. Lauren Kaori Gurley, "Amazon Drivers Are Instructed to Drive Recklessly to Meet Delivery Quotas," *Vice*, May 6, 2021, https://www.vice.com/en/article/xgxx54/amazon-drivers-are-instructed-to-drive-recklessly-to-meet-delivery-quotas.

42. Belzer, *Sweatshops on Wheels*.

43. Willis, *Learning to Labor*, 3.

44. Ewick and Silbey, "Narrating Social Structure," 1331.

45. Ashley T. Rubin, "The Consequences of Prisoners' Micro-Resistance," *Law & Social Inquiry* 42, no. 1 (2017): 138–62.

46. Torin Monahan, "The Right to Hide? Anti-Surveillance Camouflage and The Aestheticization of Resistance," *Communication and Critical/Cultural Studies* 12, no. 2 (2015): 159–78.

47. Patricia Ewick and Susan S. Silbey, "Conformity, Contestation, and Resistance: An Account of Legal Consciousness," *New England Law Review* 26 (1991): 748.

48. Ewick and Silbey, "Conformity, Contestation, and Resistance," 749.

Chapter 7. RoboTruckers: The Double Threat of AI for Low-Wage Work

1. Daron Acemoglu and David Autor, "Skills, Tasks and Technologies: Implications for Employment and Earnings," in *Handbook of Labor Economics* 4 (2011): 1043–1171.

2. David H. Autor, Frank Levy, and Richard J. Murnane, "The Skill Content of Recent Technological Change: An Empirical Exploration," *Quarterly Journal of Economics* 118, no. 4 (2003): 1286.

3. Autor, Levy, and Murnane, 1283.

4. Carl Benedikt Frey and Michael A. Osborne, "The Future of Employment: How Susceptible Are Jobs to Computerisation?" *Technological Forecasting and Social Change* 114 (January 2017): 254–80. (Despite its official publication date, I refer to the Frey and Osborne paper as a "2013 paper" because it became widely cited after being released as a working paper in 2013.)

5. Council of Economic Advisors, "Economic Report of the President," February 2016, https://obamawhitehouse.archives.gov/sites/default/files/docs/ERP_2016_Book_Complete%20JA.pdf, pp. 238–39.

6. "Artificial Intelligence, Automation, and the Economy," Report of the Executive Office of the President, December 2016, https://obamawhitehouse.archives.gov/sites/whitehouse.gov/files/documents/Artificial-Intelligence-Automation-Economy.pdf, pp. 15–17.

7. Frey and Osborne, "Future of Employment."

8. Conor Dougherty, "Self-Driving Trucks May Be Closer Than They Appear," *New York Times*, November 13, 2017, https://www.nytimes.com/2017/11/13/business/self-driving-trucks.html.

9. International Transport Forum, "Managing the Transition to Driverless Road Freight Transport," May 31, 2017, https://www.itf-oecd.org/sites/default/files/docs/managing-transition-driverless-road-freight-transport.pdf.

10. Evan Halper, "The Driverless Revolution May Exact a Political Price," *Los Angeles Times*, November 21, 2017, https://www.latimes.com/politics/la-na-pol-self-driving-politics-20171121-story.html.

11. Tim Dickinson, "Death of the American Trucker," *Rolling Stone*, January 2, 2018, https://www.rollingstone.com/politics/politics-features/death-of-the-american-trucker-253712/.

12. Jack Nerad, "Trucking Industry Seeks to Avoid 'Robot Apocalypse,'" *Trucks.com*, May 9, 2018, https://www.trucks.com/2018/05/09/trucking-industry-robot-apocalypse/.

13. Brad Templeton, "Starsky Robotics Shuts Down and Worries Everybody Else Will Also Fail in Robotic Trucks," *Forbes*, April 2, 2020, https://www.forbes.com/sites/bradtempleton/2020/04/02/starsky-robotics-shuts-down-and-worries-everybody-else-will-also-fail-in-robotic-trucks/.

14. Dougherty, "Self-Driving Trucks May Be Closer than They Appear."

15. James Jaillet, "Forecasters: Trucking's Autonomous Revolution Is Nigh, and It May Be Prickly," *Overdrive*, October 2, 2016, https://www.overdriveonline.com/forecasters-truckings-autonomous-revolution-is-nigh-and-it-may-be-prickly/.

16. Dougherty, "Self-Driving Trucks May Be Closer than They Appear."

17. Joseph Stromberg, "This Is the First Licensed Self-Driving Truck. There Will Be Many More," *Vox*, May 6, 2015, https://www.vox.com/2014/6/3/5775482/why-trucks-will-drive-themselves-before-cars-do.

18. Cameron F. Kerry and Jack Karsten, "Gauging Investment in Self-Driving Cars," *Brookings*, October 16, 2017, https://www.brookings.edu/research/gauging-investment-in-self-driving-cars/.

19. McKinsey Center for Future Mobility, "The Future of Mobility is at Our Doorstep," McKinsey Compendium 2019/2020, https://www.mckinsey.com/~/media/McKinsey/Industries/Automotive%20and%20Assembly/Our%20Insights/The%20future%20of%20mobility%20is%20at%20our%20doorstep/The-future-of-mobility-is-at-our-doorstep.ashx.

20. Ed Garsten, "Sharp Growth in Autonomous Car Market Value Predicted but May Be Stalled by Rise in Consumer Fear," *Forbes*, August 13, 2018, https://www.forbes.com/sites/edgarsten/2018/08/13/sharp-growth-in-autonomous-car-market-value-predicted-but-may-be-stalled-by-rise-in-consumer-fear/#4e1144b4617c.

21. Matt McFarland, "A Self-Driving Truck Just Hauled 51,744 Cans of Budweiser on a Colorado Highway," *CNN Business*, October 25, 2016, https://money.cnn.com/2016/10/25/technology/otto-budweiser-self-driving-truck/index.html.

22. Alex Davies, "Self-Driving Trucks Are Now Delivering Refrigerators," *Wired*, November 13, 2017, https://www.wired.com/story/embark-self-driving-truck-deliveries/.

23. Joon Ian Wong, "A Fleet of Trucks Just Drove Themselves across Europe," *Quartz*, April 6, 2016, https://qz.com/656104/a-fleet-of-trucks-just-drove-themselves-across-europe/.

24. "Otto's Anthony Levandowski on Autonomous Benefits," YouTube video of panel at American Trucking Associations' 2016 Management Conference and Exhibition, October 18, 2016, https://www.youtube.com/watch?v=Eq7W8FI76sA.

25. For example, in Erik Brynjolffsson and colleagues' analysis of the occupational impacts of machine learning (ML), they clarify that their analysis "focuses on technical feasibility. It is silent on the economic, organizational, legal, cultural, and societal factors influencing ML adoption." Erik Brynjolfsson, Tom Mitchell, and Daniel Rock, "What Can Machines Learn and What Does It Mean for Occupations and the Economy?" *AEA Papers and Proceedings* 108 (2018): 45. Similarly, Frey and Osborne's study points out that the jobs they consider to be at risk are due to the *technological capabilities* of machines, and how they match up with the task content of the job—but that "several additional factors," including the cost of human labor, social resistance, and regulatory concerns, will influence how many jobs are *actually* replaced. Frey and Osborne, "Future of Employment." Other researchers have expressed related doubts about some of the more alarmist forecasts of AI's impact on jobs, pointing out both the technical limitations of AI technologies to date, as well as social and organizational factors that are likely to temper its effects. Leslie Willcocks, "Robo-Apocalypse Cancelled? Reframing the Automation and Future of Work Debate," *Journal of Information Technology* 35, no. 4 (2020): 286–302; Meredith Broussard, *Artificial Unintelligence: How Computers Misunderstand the World* (Cambridge: MA: MIT Press, 2018).

26. Steve Viscelli, "Driverless? Autonomous Trucks and the Future of the American Trucker," Report of the Center for Labor Research and Education, University of California, Berkeley, and Working Partnerships USA, September 2018, http://driverlessreport.org/files/driverless.pdf, pp. 16–17.

27. Aleksandr Yankelevich et al., "Preparing the Workforce for Automated Vehicles," report submitted to The American Center for Mobility, July 30, 2018, https://comartsci.msu.edu/sites/default/files/documents/MSU-TTI-Preparing-Workforce-for-AVs-and-Truck-Platooning-Reports%20.pdf.

28. Michael Polanyi, *The Tacit Dimension* (Chicago: University of Chicago Press, 1966).

29. Polanyi, *Tacit Dimension*, quoted in Autor, Levy, and Murnane, "Skill Content of Recent Technological Change," 1283.

30. The need for coordination between human and machine "drivers" prompts interesting ethical questions. One option might be to program autonomous vehicles to break or bend rules more often, like human drivers do—though of course, programming a vehicle to break the law raises a host of important questions about liability and ethics. Others suggest creating consistency by making human drivers drive more like robots do—for example, by using technological solutions like speed limiters to keep humans from breaking rules as frequently as they do. See Sven Nyholm

and Jilles Smids, "Automated Cars Meet Human Drivers: Responsible Human-Robot Coordination and the Ethics of Mixed Traffic," *Ethics and Information Technology* 22, no. 4 (2018): 335–44; Oskar Juhlin, "Traffic Behaviour as Social Interaction—Implications for the Design of Artificial Drivers," in *Proceedings of the 6th World Congress on Intelligent Transport Systems* (1999): 8–12; Fridulv Sagberg et al., "A Review of Research on Driving Styles and Road Safety," *Human Factors: The Journal of the Human Factors and Ergonomics Society* 57, no. 7 (2015): 1248–75.

31. Tom Krisher, "The Toughest Challenge for Self-Driving Cars? Human Drivers," *AP News*, May 11, 2017, https://apnews.com/65e1c800ff0c4532abbdabc134e5f9b8.

32. Brynjolfsson, Mitchell, and Rock, "What Can Machines Learn."

33. James Pethokoukis, "What the Story of ATMs and Bank Tellers Reveals About the 'Rise of the Robots' and Jobs," American Enterprise Institute, June 6, 2016, https://www.aei.org/economics/what-atms-bank-tellers-rise-robots-and-jobs/.

34. John Gerzema, "Breaking down the Barriers of Self-Driving Cars," *Forbes*, November 16, 2017, https://www.forbes.com/sites/forbesagencycouncil/2017/11/16/breaking-down-the-barriers-of-self-driving-cars/?sh=6e7be4d22ef1.

35. Robert Ferris, "There Is a Ton of People Who Still Don't Want to Ride in Self-Driving Cars, Says Survey," *CNBC*, August 24, 2017, https://www.cnbc.com/2017/08/24/consumers-still-anxious-about-autonomous-cars-says-gartner.html.

36. Keith Naughton, "Most Americans Wary of Self-Driving Tech, Don't Want Robo Cars," *Bloomberg*, September 6, 2017, https://www.bloomberg.com/news/articles/2017-09-06/most-americans-wary-of-self-driving-tech-don-t-want-robo-cars.

37. Dana Hull, Mark Bergen, and Polly Mosendz, "The Uber Crash Is the Nightmare the Driverless World Feared but Expected," *Bloomberg*, March 19, 2018, https://www.bloomberg.com/news/articles/2018-03-19/uber-crash-is-nightmare-the-driverless-world-feared-but-expected.

38. Felicia Fonseca and Tom Krisher, "Self-Driving Vehicle Hits, Kills Walker in Arizona," *Columbus Dispatch*, March 20, 2018, http://www.pressreader.com/usa/the-columbus-dispatch/20180320/281513636687475, p. A4.

39. Alan Levin and Ryan Beene, "Tesla Model X in California Crash Sped Up Prior to Impact," *Bloomberg*, June 7, 2018, https://www.bloomberg.com/news/articles/2018-06-07/tesla-model-x-in-california-crash-sped-up-seconds-before-impact.

40. "Teamsters Union Perspective on Automation Featured in *Rolling Stone*," *International Brotherhood of Teamsters*, January 2, 2018, https://teamster.org/2018/01/teamsters-union-perspective-automation-featured-rolling-stone/.

41. Ryan Hagemann, "Senseless Government Rules Could Cripple the Robo-Car Revolution," *Wired*, May 1, 2017, https://www.wired.com/2017/05/senseless-government-rules-cripple-robo-car-revolution/.

42. Erik Brynjolfsson and Tom Mitchell, "What Can Machine Learning Do? Workforce Implications," *Science* 358, no. 6370 (December 22, 2017): 1534. Brynjolfsson and Mitchell note that legal constraints like these are "subtle and difficult to identify, and they can create considerable inertia" to the adoption and diffusion of new technologies.

43. David Perlman et al., "Review of the Federal Motor Carrier Safety Regulations for Automated Commercial Vehicles: Preliminary Assessment of Interpretation and Enforcement Challenges, Questions, and Gaps," summary report prepared for Intelligent Transportation Systems Joint Program Office and Federal Motor Carrier Safety Administration, Report no. FMCSA-RRT-17–013, March 2018, https://rosap.ntl.bts.gov/view/dot/35426.

44. 49 C.F.R. § 392.9.

45. Bryant Walker Smith, "A Legal Perspective on Three Misconceptions in Vehicle Automation," in *Road Vehicle Automation*, eds. Gereon Meyer and Sven Beiker (New York: Springer, 2014): 85–91.

46. Bibb v. Navajo Freight Lines, Inc., 359 U.S. 520 (1959); Raymond Motor Transportation, Inc. v. Rice, 434 U.S. 429 (1978).

47. Jack Denton, "Are the Teamsters Trying to Kill Driverless Tech, or Save the Truck Drivers?" *Pacific Standard*, April 27, 2018, https://psmag.com/economics/trucking-teamsters -driverless-tech; Todd Dills, "Autonomous Revolution May Be Nigh, but It's Coming for Brokers Faster Than Operators," *Overdrive*, May 28, 2018, https://www.overdriveonline.com/autonomous -revolution-may-be-nigh-but-its-coming-for-brokers-faster-than-drivers/; Karlyn D. Stanley, Michelle Grisé, and James M. Anderson, "Autonomous Vehicles and the Future of Auto Insurance," RAND (report), 2020, https://www.rand.org/pubs/research_reports/RRA878-1.html.

48. Jake Goldenfein, Deirdre Mulligan, Helen Nissenbaum, and Wendy Ju take a broad view of the handoff in their exploration of competing visions of autonomous vehicle systems, noting that the handoffs of functionalities are not merely technical or mechanical in nature, but also entail broad and complex ethical and social reconfigurations of values—for example, influencing our notions of responsibility, privacy, and ownership. Jake Goldenfein et al., "Through the Handoff Lens: Competing Visions of Autonomous Futures," *Berkeley Technology Law Journal* 35, no. 3 (2020): 835–910.

49. John Maynard Keynes, "Economic Possibilities for Our Grandchildren," in *Essays in Persuasion* (London: Harcourt Brace, 1932), 367.

50. Antonio Regalado, "When Machines Do Your Job," *MIT Technology Review*, July 11, 2012, https://www.technologyreview.com/s/428429/when-machines-do-your-job/.

51. Robert B. Reich, *Saving Capitalism: For the Many, Not the Few* (New York: Alfred A. Knopf, 2016), 215.

52. Ross Miller, "AP's 'Robot Journalists' Are Writing Their Own Stories Now," *Verge*, January 29, 2015, https://www.theverge.com/2015/1/29/7939067/ap-journalism-automation-robots -financial-reporting.

53. Paul Colford, "A Leap Forward in Quarterly Earnings Stories," *Associated Press Definitive Source*, June 30, 2014, https://blog.ap.org/announcements/a-leap-forward-in-quarterly-earnings -stories.

54. Alex Wilhelm, "Bring on the Blogging Robots," *TechCrunch*, July 1, 2014, https:// techcrunch.com/2014/07/01/bring-on-the-blogging-robots/.

55. Meredith Broussard et al., "Artificial Intelligence and Journalism," *Journalism & Mass Communication Quarterly* 96, no. 3 (2019): 673–95; Angèle Christin, "Counting Clicks: Quantification and Variation in Web Journalism in the United States and France," *American Journal of Sociology* 123, no. 5 (2018): 1382–1415.

56. Rob Stumpf, "Tesla Admits Current 'Full Self-Driving Beta' Will Always Be a Level 2 System: Emails," *Drive*, March 8, 2021, https://www.thedrive.com/tech/39647/tesla-admits-current -full-self-driving-beta-will-always-be-a-level-2-system-emails.

57. Carolyn Said, "Robot Cars May Kill Jobs, but Will They Create Them Too?" *San Francisco Chronicle*, December 8, 2017, https://www.sfchronicle.com/news/article/Robot-cars-may -kill-jobs-but-will-they-create-12410820.php. The American Transportation Research Institute similarly forecasts that drivers might complete logistical or administrative tasks, or rest, at a high enough level of vehicle autonomy. Jeffrey Short and Dan Murray, "Identifying Autonomous Vehicle Technology Impacts on the Trucking Industry," American Transportation Research Institute (report), November 2016, https://truckingresearch.org/wp-content/uploads/2016/11/ATRI -Autonomous-Vehicle-Impacts-11-2016.pdf.

58. April Glaser, "Autonomous Tech Could Make Driving Semi-Trucks Even Less Fun," *Wired*, April 27, 2016, https://www.wired.com/2016/04/autonomous-tech-make-driving-semi-trucks -even-less-fun/.

59. David H. Freedman, "Self-Driving Trucks: 10 Breakthrough Technologies 2017," *MIT Technology Review*, February 22, 2017, archived at https://web.archive.org/web/20200101172733

/https://www.technologyreview.com/s/603493/10-breakthrough-technologies-2017-self-driving-trucks/.

60. Joost C. F. de Winter et al., "Effects of Adaptive Cruise Control and Highly Automated Driving on Workload and Situation Awareness: A Review of the Empirical Evidence," *Transportation Research Part F* 27 (2014): 208.

61. Stephen M. Casner, Edwin L. Hutchins, and Don Norman, "The Challenges of Partially Automated Driving," *Communications of the ACM* 59, no. 5 (May 2016): 73.

62. Casner, Hutchins, and Norman, 73.

63. Myra Blanco et al., "Human Factors Evaluation of Level 2 and Level 3 Automated Driving Concepts," National Highway Traffic Safety Administration Report no. DOT HS 812 182, August 2015, https://www.nhtsa.gov/sites/nhtsa.gov/files/812182_humanfactorseval-l2l3-automdrivingconcepts.pdf, p. 104.

64. Lisanne Bainbridge, "Ironies of Automation," *Automatica* 19, no. 6 (1983): 775.

65. Stephen M. Casner et al., "The Retention of Manual Flying Skills in the Automated Cockpit," *Human Factors: The Journal of the Human Factors and Ergonomics Society* 56, no. 8 (2014): 1506–16.

66. United States Department of Transportation Office of Inspector General, "Enhanced FAA Oversight Could Reduce Hazards Associated with Increased Use of Flight Deck Automation," Report No. AV-2016-013, January 7, 2016, https://www.oig.dot.gov/sites/default/files/FAA%20Flight%20Decek%20Automation_Final%20Report%5E1-7-16.pdf.

67. P. A. Hancock, "Some Pitfalls in the Promises of Automated and Autonomous Vehicles," *Ergonomics* 62, no. 4 (2019): 514–20.

68. Moritz Körber et al., "Vigilance Decrement and Passive Fatigue Caused by Monotony in Automated Driving," *Procedia Manufacturing* 3 (2015): 2403–09.

69. Casner, Hutchins, and Norman, "Challenges of Partially Automated Driving," 74.

70. Dyani J. Saxby et al., "Active and Passive Fatigue in Simulated Driving: Discriminating Styles of Workload Regulation and Their Safety Impacts," *Journal of Experimental Psychology: Applied* 19, no. 4 (2013): 287–300.

71. Hancock, "Some Pitfalls."

72. N. H. Mackworth, "The Breakdown of Vigilance during Prolonged Visual Search," *Quarterly Journal of Experimental Psychology* 1, no. 1 (1948): 6–21.

73. Eric T. Greenlee, Patricia R. DeLucia, and David C. Newton, "Driver Vigilance in Automated Vehicles: Hazard Detection Failures Are a Matter of Time," *Human Factors: The Journal of the Human Factors and Ergonomics Society* 60, no. 4 (2018): 465–76.

74. Tobias Vogelpohl et al., "Asleep at the Automated Wheel—Sleepiness and Fatigue during Highly Automated Driving," *Accident Analysis & Prevention* 126 (May 2019): 70–84.

75. A further important consideration is where responsibility lies in such systems. Even when control over a task is distributed between human workers and machines, we tend to attribute legal and moral responsibility for failures to the human alone. Madeleine Clare Elish writes that the human in machine/human systems operates as a "moral crumple zone," absorbing blame when things go awry. Elish points to the Three Mile Island nuclear accident as an example: though the meltdown was a result of a variety of interconnected failures among cooling pipes, clogged filters, confusing control interfaces, and human operators, press accounts largely attributed the disaster to "human error," ignoring the mechanical components that failed to operate as planned. Similar issues arise in aviation: even when autopilot systems malfunction or communicate incorrect or confusing information to human pilots, plane crashes are attributed primarily to pilots' mistakes—including their failure to recognize when the *machine* has failed. Madeleine Clare Elish, "Moral Crumple Zones: Cautionary Tales in Human-Robot Interaction," *Engaging Science, Technology, and Society* 5 (2019): 40–60.

76. Bainbridge, "Ironies of Automation," 777.

77. "Volvo Cars CEO Urges Governments and Car Industry to Share Safety-Related Traffic Data," Volvo Cars Global Media Newsroom, April 3, 2017, https://www.media.volvocars.com/global/en-gb/media/pressreleases/207164/volvo-cars-ceo-urges-governments-and-car-industry-to-share-safety-related-traffic-data.

78. Mark Fields, "Ford's Road to Full Autonomy," LinkedIn post, April 16, 2016, https://www.linkedin.com/pulse/fords-road-full-autonomy-mark-fields/.

79. Overdrive Staff, "Daimler Trucks Building Facility Dedicated to Autonomous Research, Development," *Overdrive*, June 13, 2018, https://www.overdriveonline.com/business/article/14894536/daimer-trucks-to-open-new-research-development-center.

80. John Markoff, "Google's Next Phase in Driverless Cars: No Steering Wheel or Brake Pedals," *New York Times*, May 27, 2014, https://www.nytimes.com/2014/05/28/technology/googles-next-phase-in-driverless-cars-no-brakes-or-steering-wheel.html.

81. Uber Advanced Technologies Group, "The Future of Trucking," Medium post, November 14, 2017, archived at https://web.archive.org/web/20180210230753/https://medium.com/@UberATG/the-future-of-trucking-292202cb42a2.

82. Uber Advanced Technologies Group, "The Future of Trucking: Mixed Fleets, Transfer Hubs, and More Opportunity for Truck Drivers," Medium post, February 1, 2018, archived at https://web.archive.org/web/20180201192005/https://medium.com/@UberATG/the-future-of-trucking-b3d2ea0d2db9.

83. Alexis C. Madrigal, "Could Self-Driving Trucks Be Good for Truckers?" *Atlantic*, February 1, 2018, https://www.theatlantic.com/technology/archive/2018/02/uber-says-its-self-driving-trucks-will-be-good-for-truckers/551879/. Viscelli argues for state-level wage boards to help ensure stable wages and good working conditions for truckers under this scenario. Viscelli, "Driverless?," 48.

84. Kirsten Korosec, "Uber's Self-Driving Trucks Division Is Dead, Long Live Uber Self-Driving Cars," *TechCrunch*, July 30, 2018, https://techcrunch.com/2018/07/30/ubers-self-driving-trucks-division-is-dead-long-live-uber-self-driving-cars/.

85. Cade Metz and Kate Conger, "Uber, After Years of Trying, Is Handing off Its Self-Driving Car Project," *New York Times*, December 7, 2020, https://www.nytimes.com/2020/12/07/technology/uber-self-driving-car-project.html.

86. Jamie Condliffe, "Waymo and Uber Have Reached a Settlement in Their Trade Secrets Battle," *MIT Technology Review*, February 9, 2018, https://www.technologyreview.com/the-download/610236/waymo-and-uber-have-reached-a-settlement-in-their-trade-secrets-trial/.

87. Kirsten Korosec, "This Driverless Truck Startup Is Putting Human Drivers to Work," *Fortune*, February 28, 2017, https://fortune.com/2017/02/28/starsky-self-driving-truck-startup/.

88. Tim Higgins, "Driverless Cars Still Handled by Humans-from Afar," *Wall Street Journal*, June 5, 2018, https://www.wsj.com/articles/who-does-a-driverless-car-call-when-it-needs-help-a-human-1528191000.

89. Higgins, "Driverless Cars."

90. Stefan Seltz-Axmacher, "The End of Starsky Robotics," Medium post, March 19, 2020, https://medium.com/starsky-robotics-blog/the-end-of-starsky-robotics-acb8a6a8a5f5.

91. Manfred E. Clynes and Nathan S. Kline, "Cyborgs and Space," *Astronautics*, September 1960, pp. 26–27, 76.

92. Clynes and Kline, 27.

93. Clynes and Kline, 76.

94. Toby Freedman, "'Robot Man' Not Space Solution," *Miami News*, May 3, 1964, p. 10, archived at https://perma.cc/JS9T-3GMK.

95. Legal scholar Andrea Matwyshyn raises concerns that arise due to the "platformization" of the human body as "computer code and human corpora blend," including concerns about privacy,

liability, and autonomy. Andrea M. Matwyshyn, "The Internet of Bodies," *William & Mary Law Review* 61, no. 1 (2019): 81.

96. James Brooks, "Cyborgs at Work: Employees Get Microchipped," *San Francisco Chronicle,* April 3, 2017, https://www.sfgate.com/business/article/Cyborgs-at-work-Employees-get-microchipped-11047084.php.

97. Humanyze, 2011, https://www.humanyze.com/.

98. Heather Kelly, "Amazon's Idea for Employee-Tracking Wearables Raises Concerns," *CNN Business*, February 2, 2018, https://money.cnn.com/2018/02/02/technology/amazon-employee-tracker/index.html.

99. Supriya Rai, "Tech Mahindra's Moodometer Gauges Employees' Mood to Uplift Work Environment." *Dataquest*, August 10, 2018, https://www.dqindia.com/tech-mahindras-moodometer-gauges-employees-mood-uplift-work-environment/.

100. Bryan Menegus, "Amazon Imagines Future Employee Surveillance with Patent Application for AR Goggles," *Gizmodo*, August 2, 2018, https://gizmodo.com/amazon-imagines-future-employee-surveillance-with-paten-1828051062.

101. Heather J. Hicks, *The Culture of Soft Work: Labor, Gender, and Race in Postmodern American Narrative* (New York: Palgrave Macmillan, 2009), 136.

102. Mark Schremmer, "New Gadgets Move Us Closer to a World of 'RoboTruckers,'" *Land Line*, January 11, 2017, https://landline.media/new-gadgets-move-us-closer-to-a-world-of-robotruckers/.

103. Stephen Chen, "'Forget the Facebook Leak': China Is Mining Data Directly from Workers' Brains on an Industrial Scale," *South China Morning Post*, April 29, 2018, https://www.scmp.com/news/china/society/article/2143899/forget-facebook-leak-china-mining-data-directly-workers-brains.

104. Marlin Caddell, "Fatigue's Fast Track—Body Language: Wearable Monitor Makers See Inroads in Trucking Industry Detecting Fatigue," *Overdrive*, November 2, 2017, https://www.overdriveonline.com/fatigues-fast-track-body-language-wearable-monitor-makers-see-inroads-in-trucking-industry-detecting-fatigue/.

105. Caddell, "Fatigue's Fast Track—Body Language."

106. Tim Collins, "Steer Device Delivers Shocks to Drivers Who Fall Asleep," *Daily Mail*, July 31, 2017, https://www.dailymail.co.uk/sciencetech/article-4746076/Steer-delivers-shocks-drivers-fall-asleep.html.

107. Samuel Barradas, "Smile! Facial Recognition Software Is Coming to a Truck Near You," *Truckers Report*, 2016, https://www.thetruckersreport.com/smile-facial-recognition-software-coming-truck-near/.

108. "Real-Time Driver Fatigue Monitoring System: *Guardian*," Seeing Machines, https://www.seeingmachines.com/guardian/guardian/.

109. Todd Dills, "One Year In: Lytx ActiveVision Camera System," *Overdrive*, September 21, 2016, https://www.overdriveonline.com/one-year-in-lytx-activevision-camera-system/.

110. Todd Dills, "Camera Systems Taking a Lead in Data Harvest," *Overdrive*, April 21, 2016, https://www.overdriveonline.com/business/article/14890160/camera-systems-taking-a-lead-in-data-harvest; Todd Dills, "Dashcam Video Monitoring Platform with 'Artificial Intelligence' Bonus Aims at Driver-Rewards Focus," *Overdrive*, September 16, 2016, https://www.overdriveonline.com/dashcam-video-monitoring-platform-with-artificial-intelligence-bonus-aims-at-driver-rewards-focus/.

111. James Jaillet, "Fatigue's Fast Track—Regulations: Are Fatigue-Monitoring Systems Destined to Upend Hours-Based Regulations?" *Overdrive*, September 23, 2019, https://www.overdriveonline.com/fatigues-fast-track-regulations-are-fatigue-monitoring-systems-destined-to-upend-hours-based-regulations/.

112. Todd Dills, "Fatigue's Fast Track—Dashcam Monitoring: Private Eyes, Watching You," *Overdrive*, December 18, 2018, https://www.overdriveonline.com/regulations/article/14893111 /fatigues-fast-track-dashcam-monitoring-private-eyes-watching-you.

113. Caddell, "Fatigue's Fast Track—Body Language."

114. Jaillet, "Fatigue's Fast Track—Regulations."

115. Max Heine, "A Potential Role for Cameras and Wearables in Regulating Drowsy Driving," *Overdrive*, November 26, 2017, https://www.overdriveonline.com/fatigue-monitoring-eventual -replacement-for-hours-of-service/.

116. Application of Pronto.AI, Inc. for an Exemption from 49 CFR 395.3(a)(2) and 395.3(a)(3) for Motor Carriers Using Certain Advanced Driver Assistance Systems, April 20, 2020, FMCSA-2020-0116-0004, https://www.regulations.gov/document/FMCSA-2020-0116-0004.

117. Todd Dills, "Potential for Fatigue Detection Points to Intensive Enforcement," *Overdrive*, April 24, 2016, https://www.overdriveonline.com/tomorrows-trucker-potential-for-fatigue -detection-points-to-intensive-enforcement/.

118. Samuel Barradas, "AAA Calls for Driver-Facing Cameras, More Tech in Trucks," *Truckers Report*, 2017, https://www.thetruckersreport.com/aaa-calls-driver-facing-cameras-tech -trucks/.

119. Samuel Barradas, "Truckers Targeted by Facial Recognition Artificial Intelligence," *Truckers Report*, 2018, https://www.thetruckersreport.com/truckers-targeted-facial-recognition -artificial-intelligence/.

120. Schremmer, "New Gadgets Move Us Closer to a World of 'RoboTruckers.'"

121. Thomas B. Sheridan, "Big Brother as Driver: New Demands and Problems for the Man at the Wheel," *Human Factors: The Journal of the Human Factors and Ergonomics Society* 12, no. 1 (1970): 95.

122. Elizabeth Dwoskin and Mike Ramsey, "Car Makers Test Technology to Make You Pay Attention to the Road," *Wall Street Journal*, March 11, 2016, https://www.wsj.com/articles/car-tech -that-watches-how-you-drive-1457692201; Eric A. Taub, "Sleepy Behind the Wheel? Some Cars Can Tell," *New York Times*, March 16, 2017, https://www.nytimes.com/2017/03/16/automobiles /wheels/drowsy-driving-technology.html.

123. Alex Davies, "Audi's New A8 Shows How Robocars Can Work with Humans," *Wired*, July 11, 2017, https://www.wired.com/story/self-driving-car-user-interface/.

124. Heavy Duty Trucking Staff, "ZF Details Research on Passenger Monitoring System," *Truckinginfo*, October 16, 2018, https://www.truckinginfo.com/316828/zf-details-research-on -passenger-monitoring-system.

125. Chris Tennant and Jack Stilgoe, "The Attachments of 'Autonomous' Vehicles," *Social Studies of Science* 51, no. 6 (2021): 849.

126. National Highway Traffic Safety Administration, "Consumer Advisory: NHTSA Deems 'Autopilot Buddy' Product Unsafe, June 19, 2018, https://www.nhtsa.gov/press-releases/consumer -advisory-nhtsa-deems-autopilot-buddy-product-unsafe.

127. Dave Vanderwerp, "It's Not Just Tesla: All Other Driver-Assist Systems Work without Drivers, Too," *Car and Driver*, August 11, 2021, https://www.caranddriver.com/news/a37260363 /driver-assist-systems-tested/.

128. National Transportation Safety Board, "Accident Report: Collision Between a Car Operating with Automated Vehicle Control Systems and a Tractor-Semitrailer Truck Near Williston, Florida," NTSB/HAR-17/02 PB2017–102600, May 7, 2016, https://www.ntsb.gov/investigations /AccidentReports/Reports/HAR1702.pdf.

129. Sean O'Kane, "Tesla Starts Using In-Car Camera for Autopilot Driver Monitoring," *Verge*, May 27, 2021, https://www.theverge.com/2021/5/27/22457430/tesla-in-car-camera-driver -monitoring-system.

130. Timothy B. Lee, "Legislation Would Mandate Driver-Monitoring Tech in Every Car," *Ars Technica*, April 27, 2021, https://arstechnica.com/tech-policy/2021/04/legislation-would-mandate-driver-monitoring-tech-in-every-car/.

Chapter 8. Technology, Enforcement, and Apparent Order

1. Harvey Molotch, *Against Security: How We Go Wrong at Airports, Subways, and Other Sites of Ambiguous Danger* (Princeton: Princeton University Press, 2014), 215–16. James C. Scott makes a related observation in the context of governance and state planning, observing that "[f]ormal order . . . is always and to some considerable degree parasitic on informal processes, which the formal scheme does not recognize, without which it could not exist, and which it alone cannot create or maintain." James C. Scott, *Seeing Like a State: How Certain Schemes to Improve the Human Condition Have Failed* (New Haven: Yale University Press, 1998), 310.

2. Belzer, "Building Back Better in Commercial Road Transport."

3. Viscelli, "Driverless?," 48.

4. Bringing truckers under the ambit of fair labor laws is likely necessary but insufficient; as Viscelli points out, truckers (like many workers) are frequently misclassified as independent contractors rather than employees of firms, further attenuating their labor rights. Viscelli, *Big Rig*.

5. Rediet Abebe et al., "Roles for Computing in Social Change," in *Proceedings of the ACM Conference on Fairness, Accountability, and Transparency (FAccT)* (2020), 252–60. In this article, my colleagues and I put forth a menu of ways in which computation, in concert with other forms of knowledge, can be put to use in the service of fundamental social change.

6. Abebe et al.

7. Chelsea Barabas et al., "Studying Up: Reorienting the Study of Algorithmic Fairness Around Issues of Power," in *Proceedings of the ACM Conference on Fairness, Accountability, and Transparency (FAccT)* (2020), 167–76. Barabas and her colleagues derive the notion of studying up in data science from earlier calls to do so in anthropology. See Laura Nader, "Up the Anthropologist: Perspectives Gained from Studying Up," in *Reinventing Anthropology*, ed. Dell Hymes (New York: Pantheon Books, 1972), 284–311.

8. Mimi Onuoha, "When Proof Is Not Enough," *FiveThirtyEight*, July 1, 2020, https://fivethirtyeight.com/features/when-proof-is-not-enough/.

9. Abebe et al., "Roles for Computing in Social Change."

10. Alexandra Mateescu and Madeleine Clare Elish, *AI in Context: The Labor of Integrating New Technologies*, Data & Society Research Institute (report), 2019, https://datasociety.net/wp-content/uploads/2019/01/DataandSociety_AIinContext.pdf.

11. Blader, Gartenberg, and Prat, "Contingent Effect of Management Practices."

12. Moradi and Levy, "Future of Work in the Age of AI."

Appendix A. Studying Surveillance

1. Trevor J. Pinch and Wiebe E. Bijker, "The Social Construction of Facts and Artefacts: Or How the Sociology of Science and the Sociology of Technology Might Benefit Each Other," *Social Studies of Science* 14, no. 3 (1984): 399–441.

2. Christine Hine, "Multi-Sited Ethnography as a Middle Range Methodology for Contemporary STS," *Science, Technology, & Human Values* 32, no. 6 (2007): 652–71; George E. Marcus, "Ethnography In/of the World System: The Emergence of Multi-Sited Ethnography," *Annual Review of Anthropology* 24 (1995): 95–117.

3. Viscelli, *Big Rig*; Balay, *Semi Queer*.

4. There were a few exceptions to this rule. One driver and I spoke on the phone multiple times and also exchanged emails, including photographs he took in the course of his work. In addition, I encountered a couple of drivers more than once on different visits to a truck stop; drivers (especially those with regular routes) often develop patterns of patronage at certain truck stops, based on personal preferences, social connections, loyalty programs, and/or company rules about preferred fuel vendors. In addition, some of my relationships with non-driver participants (like trucking firm managers) could be more sustained.

5. In addition to reading many of these documents manually, I separately conducted a computational text analysis of 3,569 public comments submitted to federal agencies about electronic monitoring in trucking. That analysis focuses on the different thematic frames that undergirded different parties' viewpoints. The analysis shows that comments submitted by individuals (primarily truck drivers) were more likely to frame the electronic monitoring debate in the context of broader logistical problems plaguing the industry, like detention time and parking shortages. Meanwhile, organizational stakeholders (like trucking firms and interest groups) were more likely to frame their comments in terms of cost/benefit analysis and technical standards. Levy and Franklin, "Driving Regulation."

6. Janice M. Morse et al., "Verification Strategies for Establishing Reliability and Validity in Qualitative Research," *International Journal of Qualitative Methods* 1, no. 2 (2002): 13–22.

7. Torin Monahan and Jill A. Fisher, "Benefits of 'Observer Effects': Lessons from the Field," *Qualitative Research* 10, no. 3 (2010): 357–76.

8. Michael W. Morris et al., "Views from Inside and Outside: Integrating Emic and Etic Insights about Culture and Justice Judgment," *Academy of Management Review* 24, no. 4 (1999): 781–96.

9. Jenna Burrell, "User Agency in the Middle Range: Rumors and the Reinvention of the Internet in Accra, Ghana," *Science, Technology, & Human Values* 36, no. 2 (2011): 155.

10. Monahan and Fisher, "Benefits of 'Observer Effects.'"

11. Colin Jerolmack and Shamus Khan, "Talk Is Cheap: Ethnography and the Attitudinal Fallacy," *Sociological Methods & Research* 43, no. 2 (2014): 178–209.

12. Clive Seale, "Quality in Qualitative Research," *Qualitative Inquiry* 5, no. 4 (1999): 465–78.

13. Charlotte Aull Davies, *Reflexive Ethnography: A Guide to Researching Selves and Others* (Routledge, 2008).

BIBLIOGRAPHY

"63% of Drivers are Detained More than 3 Hours Per Stop." *DAT Freight & Analytics*, July 12, 2016. https://www.dat.com/blog/post/63-of-drivers-are-detained-more-than-3-hours-per-stop.

Abebe, Rediet, Solon Barocas, Jon Kleinberg, Karen Levy, Manish Raghavan, and David G. Robinson. "Roles for Computing in Social Change." In *Proceedings of the ACM Conference on Fairness, Accountability, and Transparency (FAccT)*, 252–60. New York: Association for Computing Machinery, 2020.

Abramson, Howard. "The Trucks Are Killing Us." *New York Times*, August 22, 2015. https://www.nytimes.com/2015/08/22/opinion/the-trucks-are-killing-us.html.

Abril, Patricia Sánchez, Avner Levin, and Alissa Del Riego. "Blurred Boundaries: Social Media Privacy and the Twenty-First-Century Employee." *American Business Law Journal* 49, no. 1 (2012): 63–124.

Acemoglu, Daron, and David Autor. "Skills, Tasks and Technologies: Implications for Employment and Earnings." In *Handbook of Labor Economics* 4 (2011): 1043–1171.

Ajunwa, Ifeoma, Kate Crawford, and Jason Schultz. "Limitless Worker Surveillance." *California Law Review* 105, no. 3 (2017): 735–76.

Allyn, Bobby. "Your Boss is Watching You: Work-From-Home Boom Leads to More Surveillance." *NPR*, May 13, 2020. https://www.npr.org/2020/05/13/854014403/your-boss-is-watching-you-work-from-home-boom-leads-to-more-surveillance.

American Trucking Associations. "Economics and Industry Data," 2020. Archived at https://web.archive.org/web/20210424163909/https://www.trucking.org/economics-and-industry-data.

American Trucking Associations. "Reports, Trends & Statistics," 2019. Archived at https://web.archive.org/web/20200128011852/https://www.trucking.org/News_and_Information_Reports_Industry_Data.aspx.

———. "Driver Shortage Update 2021," October 25, 2021. https://www.trucking.org/sites/default/files/2021-10/ATA%20Driver%20Shortage%20Report%202021%20Executive%20Summary.FINAL_.pdf.

Anderson, Ross. "On the Security of Digital Tachographs." In *Proceedings of ESORICS 98: 5th European Symposium on Research in Computer Security*, 111–25. Berlin: Springer, 1998.

Anteby, Michel. *Moral Gray Zones: Side-Production, Identity, and Regulation in an Aeronautic Plant*. Princeton: Princeton University Press, 2008.

"Artificial Intelligence, Automation, and the Economy." Report of the Executive Office of the President, December 2016. https://obamawhitehouse.archives.gov/sites/whitehouse.gov/files/documents/Artificial-Intelligence-Automation-Economy.pdf.

Ashforth, Blake E., and Glen E. Kreiner. "'How Can You Do It?': Dirty Work and the Challenge of Constructing a Positive Identity." *Academy of Management Review* 24, no. 3 (1999): 413–34.

Associated Press. "Most Truckers End 11-Day Strike." *New York Times*, February 12, 1974. https://www.nytimes.com/1974/02/12/archives/most-truckers-end-11day-strike-traffic-called-nearly-normal-despite.html.

Autor, David H., Frank Levy, and Richard J. Murnane. "The Skill Content of Recent Technological Change: An Empirical Exploration." *Quarterly Journal of Economics* 118, no. 4 (2003):
1279–1333.

Bailey, Diane E., Paul M. Leonardi, and Stephen R. Barley. "The Lure of the Virtual." *Organization
Science* 23, no. 5 (2012): 1485–1504.

Bainbridge, Lisanne. "Ironies of Automation." *Automatica* 19, no. 6 (1983): 775–79.

Balay, Anne. *Semi Queer: Inside the World of Gay, Trans, and Black Truck Drivers.* Chapel Hill,
NC: University of North Carolina Press, 2018.

Balay, Anne, and Mona Shattell. "Long-Haul Sweatshops." *New York Times*, March 9, 2016. https://
www.nytimes.com/2016/03/09/opinion/long-haul-sweatshops.html.

————"PTSD in the Driver's Seat." *Atlantic*, March 22, 2016. https://www.theatlantic.com/health
/archive/2016/03/long-haul-trucking-and-mental-health/474840/.

Balkin, Thomas J., William J. Horrey, R. Curtis Graeber, Charles A. Czeisler, and David F. Dinges.
"The Challenges and Opportunities of Technological Approaches to Fatigue Management."
Accident Analysis & Prevention 43, no. 2 (2011): 565–72.

Balko, Radley. "80 Percent of Chicago PD Dash-Cam Videos Are Missing Audio Due to 'Officer Error' or 'Intentional Destruction.'" *Washington Post*, January 29, 2016. https://www
.washingtonpost.com/news/the-watch/wp/2016/01/29/80-percent-of-chicago-pd-dash-cam
-videos-are-missing-audio-due-to-officer-error-or-intentional-destruction/.

Ball, Kirstie. "Workplace Surveillance: An Overview." *Labor History* 51, no. 1 (2010): 87–106.

Barabas, Chelsea, Colin Doyle, J. B. Rubinovitz, and Karthik Dinakar. "Studying Up: Reorienting
the Study of Algorithmic Fairness Around Issues of Power." In *Proceedings of the ACM Conference on Fairness, Accountability, and Transparency (FAccT)*, 167–76. New York: Association
for Computing Machinery, 2020.

Barley, Stephen R. "Technology as an Occasion for Structuring: Evidence from Observations
of CT Scanners and the Social Order of Radiology Departments." *Administrative Science
Quarterly* 31 (1986): 78–108.

Barradas, Samuel. "Smile! Facial Recognition Software Is Coming to a Truck Near You." *Truckers
Report*, 2016. https://www.thetruckersreport.com/smile-facial-recognition-software-coming
-truck-near/.

————. "AAA Calls for Driver-Facing Cameras, More Tech in Trucks." *Truckers Report*, 2017.
https://www.thetruckersreport.com/aaa-calls-driver-facing-cameras-tech-trucks/.

————. "Truckers Targeted by Facial Recognition Artificial Intelligence," *Truckers Report*,
2018. https://www.thetruckersreport.com/truckers-targeted-facial-recognition-artificial
-intelligence/.

Becker, Howard S. "Notes on the Concept of Commitment." *American Journal of Sociology* 66
(1960): 32–40.

Belman, Dale L., Francine Lafontaine, and Kristen A. Monaco. "Truck Drivers in the Age of
Information: Transformation Without Gain." In *Trucking in the Age of Information*, eds. Dale
Belman and Chelsea White III, 183–212. London: Routledge, 2005.

Belman, Dale L., and Kristen A. Monaco. "The Effects of Deregulation, De-Unionization, Technology, and Human Capital on the Work and Work Lives of Truck Drivers." *Industrial and
Labor Relations Review* 54, no. 2A (2001): 502–24.

Belzer, Michael. *Sweatshops on Wheels: Winners and Losers in Trucking Deregulation.* Oxford:
Oxford University Press, 2000.

————. "Building Back Better in Commercial Road Transport: Markets Require Fair Labor Standards." Presentation to Motor Carrier Safety Advisory Council of Federal Motor Carrier
Safety Administration, July 19, 2021. https://www.fmcsa.dot.gov/sites/fmcsa.dot.gov/files
/2021-07/MCSAC%20Building%20Back%20Better%20In%20Commercial%20Road%20

Transport%20-%20Markets%20Require%20Fair%20Labor%20Standards%20-%20July%20 2021_rev2.pdf.

Beninger, James R. *The Control Revolution: Technological and Economic Origins of the Information Society*. Cambridge, MA: Harvard University Press, 1986.

Bernstein, Ethan S. "Making Transparency Transparent: The Evolution of Observation in Management Theory." *Academy of Management Annals* 11, no. 1 (2017): 217–66.

Best, Amy L. *Fast Cars, Cool Rides: The Accelerating World of Youth and Their Cars*. New York: NYU Press, 2006.

Blader, Steven, Claudine Gartenberg, and Andrea Prat. "The Contingent Effect of Management Practices." *Review of Economic Studies* 87, no. 2 (2020): 721–49.

Blanco, Myra, Jon Atwood, Holland M. Vasquez, Tammy E. Trimble, Vikki L. Fitchett, Joshua Radlbeck, Gregory M. Fitch, Sheldon M. Russell, Charles A. Green, Briane Cullinane, and Justin F. Morgan. "Human Factors Evaluation of Level 2 and Level 3 Automated Driving Concepts." National Highway Traffic Safety Administration Report no. DOT HS 812 182, August 2015. https://www.nhtsa.gov/sites/nhtsa.gov/files/812182_humanfactorseval-l2l3 -automdrivingconcepts.pdf, p. 104.

Boris, Caroline, and Rebecca M. Brewster. "Managing Critical Truck Parking Case Study—Real World Insights from Truck Parking Diaries." American Transportation Research Institute (report), December 2016. https://truckingresearch.org/wp-content/uploads/2016/12/ATRI -Truck-Parking-Case-Study-Insights-12-2016.pdf.

Braver, Elisa R., Carol W. Pruesser, David F. Pruesser, Herbert M. Baum, Richard Beilock, and Robert Ulmer. "Who Violates Work Hour Rules? A Survey of Tractor-Trailer Drivers." Insurance Institute for Highway Safety (report), January 1992. http://trid.trb.org/view.aspx?id =362329.

Braverman, Harry. *Labor and Monopoly Capital: The Degradation of Work in the Twentieth Century*. New York: Monthly Review Press, 1974.

Brayne, Sarah. *Predict and Surveil: Data, Discretion, and the Future of Policing*. Oxford: Oxford University Press, 2021.

Brooks, James. "Cyborgs at Work: Employees Get Microchipped." *San Francisco Chronicle*, April 3, 2017. https://www.sfgate.com/business/article/Cyborgs-at-work-Employees-get -microchipped-11047084.php.

Broussard, Meredith. *Artificial Unintelligence: How Computers Misunderstand the World*. Cambridge, MA: MIT Press, 2018.

Broussard, Meredith, Nicholas Diakopoulos, Andrea L. Guzman, Rediet Abebe, Michel Dupagne, and Ching-Hua Chuan. "Artificial Intelligence and Journalism." *Journalism & Mass Communication Quarterly* 96, no. 3 (2019): 673–95.

Brunton, Finn, and Helen Nissenbaum. "Vernacular Resistance to Data Collection and Analysis: A Political Theory of Obfuscation." *First Monday* 16, no. 5 (2011).

Brynjolfsson, Erik, and Tom Mitchell. "What Can Machine Learning Do? Workforce Implications." *Science* 358, no. 6370 (December 22, 2017): 1530–34.

Brynjolfsson, Erik, Tom Mitchell, and Daniel Rock. "What Can Machines Learn and What Does It Mean for Occupations and the Economy?" *AEA Papers and Proceedings* 108 (2018): 43–47.

Bui, Quoctrung. "Map: The Most Common Job in Every State." NPR.org, February 5, 2015. http://www.npr.org/sections/money/2015/02/05/382664837/map-the-most-common -job-in-every-state.

Bulman-Pozen, Jessica, and David E. Pozen. "Uncivil Obedience." *Columbia Law Review* 115, no. 4 (2015): 809–72.

Burawoy, Michael. *Manufacturing Consent: Changes in the Labor Process under Monopoly Capitalism*. Chicago: University of Chicago Press, 1979.

Burks, Stephen V., Michael Belzer, Quon Kwan, Stephanie Pratt, and Sandra Shackelford. "Trucking 101: An Industry Primer." *Transportation Research Circular* no. E-C146 (December 2010). http://onlinepubs.trb.org/onlinepubs/circulars/ec146.pdf.

Burrell, Jenna. "User Agency in the Middle Range: Rumors and the Reinvention of the Internet in Accra, Ghana." *Science, Technology, & Human Values* 36, no. 2 (2011): 139–59.

Caddell, Marlin. "Fatigue's Fast Track—Body Language: Wearable Monitor Makers See Inroads in Trucking Industry Detecting Fatigue." *Overdrive*, November 2, 2017. https://www.overdriveonline.com/fatigues-fast-track-body-language-wearable-monitor-makers-see-inroads-in-trucking-industry-detecting-fatigue/.

Camden, Matthew C., Alejandra Medina-Flintsch, Jeffrey S. Hickman, Andrew M. Miller, and Richard J. Hanowski. "Leveraging Large-Truck Technology and Engineering to Realize Safety Gains: Video-Based Onboard Safety Monitoring Systems." *AAA Foundation for Traffic Safety*, September 2017. https://aaafoundation.org/wp-content/uploads/2017/11/Truck-Safety_Video-Systems.pdf.

Casner, Stephen M., Richard W. Geven, Matthias P. Recker, and Jonathan W. Schooler. "The Retention of Manual Flying Skills in the Automated Cockpit." *Human Factors: The Journal of the Human Factors and Ergonomics Society* 56, no. 8 (2014): 1506–16.

Casner, Stephen M., Edwin L. Hutchins, and Don Norman. "The Challenges of Partially Automated Driving." *Communications of the ACM* 59, no. 5 (May 2016): 70–77.

Cave, Rob. "Rise in Lorry Tachograph Tampering on UK Roads." *BBC News*, September 24, 2017. http://www.bbc.com/news/uk-41361351.

Chao, Loretta. "Trucking Makes a Comeback, but Small Operators Miss Out." *Wall Street Journal*, September 23, 2015. https://www.wsj.com/articles/trucking-makes-a-comeback-but-small-operators-miss-out-1443050680.

Chappell, Ben. *Lowrider Space: Aesthetics and Politics of Mexican American Custom Cars*. Austin, TX: University of Texas Press, 2012.

Chen, Stephen. "'Forget the Facebook Leak': China Is Mining Data Directly from Workers' Brains on an Industrial Scale." *South China Morning Post*, April 29, 2018. https://www.scmp.com/news/china/society/article/2143899/forget-facebook-leak-china-mining-data-directly-workers-brains.

Chin, Gabriel J., and Charles J. Vernon. "Reasonable but Unconstitutional: Racial Profiling and the Radical Objectivity of *Whren v. United States*." *George Washington Law Review* 83, no. 3 (2015): 882–942.

Christin, Angèle. "Counting Clicks: Quantification and Variation in Web Journalism in the United States and France." *American Journal of Sociology* 123, no. 5 (2018): 1382–1415.

"Civilian Occupations with High Fatal Work Injury Rates." United States Bureau of Labor Statistics, 2019. Archived at https://web.archive.org/web/20210126011345/https://www.bls.gov/charts/census-of-fatal-occupational-injuries/civilian-occupations-with-high-fatal-work-injury-rates.htm.

Clynes, Manfred E., and Nathan S. Kline. "Cyborgs and Space." *Astronautics*, September 1960: 26–27, 76.

Cohen, Julie E. *Between Truth and Power: The Legal Constructions of Information Capitalism*. Oxford: Oxford University Press, 2019.

Cole, Matt. "Record Number of Truckers Killed in Workplace Fatalities in 2017." *Overdrive*, December 23, 2018. https://www.overdriveonline.com/record-number-of-truckers-killed-in-workplace-fatalities-in-2017/.

Colford, Paul. "A Leap Forward in Quarterly Earnings Stories." *Associated Press Definitive Source*, June 30, 2014. https://blog.ap.org/announcements/a-leap-forward-in-quarterly-earnings-stories.

Collins, Tim. "Steer Device Delivers Shocks to Drivers Who Fall Asleep." *Daily Mail*, July 31, 2017. https://www.dailymail.co.uk/sciencetech/article-4746076/Steer-delivers-shocks-drivers-fall-asleep.html.

Condliffe, Jamie. "Waymo and Uber Have Reached a Settlement in Their Trade Secrets Battle." *MIT Technology Review*, February 9, 2018. https://www.technologyreview.com/the-download/610236/waymo-and-uber-have-reached-a-settlement-in-their-trade-secrets-trial/.

Connell, R. W. *Masculinities*. Berkeley: University of California Press, 1995.

Council of Economic Advisors. "Economic Report of the President," February 2016. https://obamawhitehouse.archives.gov/sites/default/files/docs/ERP_2016_Book_Complete%20JA.pdf.

Daipha, Phaedra. *Masters of Uncertainty: Weather Forecasters and the Quest for Ground Truth*. Chicago: University of Chicago Press, 2015.

Davies, Alex. "Audi's New A8 Shows How Robocars Can Work with Humans," *Wired*, July 11, 2017. https://www.wired.com/story/self-driving-car-user-interface/.

———. "Self-Driving Trucks Are Now Delivering Refrigerators." *Wired*, November 13, 2017. https://www.wired.com/story/embark-self-driving-truck-deliveries/.

Davies, Charlotte Aull. *Reflexive Ethnography: A Guide to Researching Selves and Others. London:* Routledge, 2008.

Day, Jennifer Cheeseman, and Andrew W. Hait. "Number of Truckers at All-Time High." United States Census Bureau, June 6, 2019. https://www.census.gov/library/stories/2019/06/america-keeps-on-trucking.html.

de Winter, Joost C.F., Riender Happee, Marieke H. Martens, and Neville A. Stanton. "Effects of Adaptive Cruise Control and Highly Automated Driving on Workload and Situation Awareness: A Review of the Empirical Evidence." *Transportation Research Part F* 27 (2014): 196–217.

Denton, Jack. "Are the Teamsters Trying to Kill Driverless Tech, or Save the Truck Drivers?" *Pacific Standard*, April 27, 2018. https://psmag.com/economics/trucking-teamsters-driverless-tech.

Derickson, Alan. *Dangerously Sleepy: Overworked Americans and the Cult of Manly Wakefulness*. Philadelphia: University of Pennsylvania Press, 2013.

Dickinson, Tim. "Death of the American Trucker." *Rolling Stone*, January 2, 2018. https://www.rollingstone.com/politics/politics-features/death-of-the-american-trucker-253712/.

Dills, Todd. "ELD Mandate: Independents' Final Straw?" *Overdrive*, April 17, 2014. https://www.overdriveonline.com/electronic-logging-devices/article/14886211/eld-mandate-independents-final-straw.

———. "Camera Systems Taking a Lead in Data Harvest." *Overdrive*, April 21, 2016. https://www.overdriveonline.com/business/article/14890160/camera-systems-taking-a-lead-in-data-harvest.

———. "Potential for Fatigue Detection Points to Intensive Enforcement." *Overdrive*, April 24, 2016. https://www.overdriveonline.com/tomorrows-trucker-potential-for-fatigue-detection-points-to-intensive-enforcement/.

———. "Stoking the Log Fires, Redux: Crash Incidence, Hours Violations and Fleet Size." *Overdrive*, July 29, 2016. https://www.overdriveonline.com/channel-19/article/14890681/stoking-the-log-fires-redux-crash-incidence-hours-violations-and-fleet-size.

———. "Dashcam Video Monitoring Platform with 'Artificial Intelligence' Bonus Aims at Driver-Rewards Focus." *Overdrive*, September 16, 2016. https://www.overdriveonline.com/dashcam-video-monitoring-platform-with-artificial-intelligence-bonus-aims-at-driver-rewards-focus/.

———. "One Year In: Lytx ActiveVision Camera System." *Overdrive*, September 21, 2016. https://www.overdriveonline.com/one-year-in-lytx-activevision-camera-system/.

———. "Older Trucks Holding Value Better since ELD Announcement." *Overdrive*, December 9, 2016. https://www.overdriveonline.com/older-trucks-holding-value-better-since-eld-announcement/.

———. "Eyes on the Prize: The Market for Pre-2000 Trucks in Light of ELD Exemption." *Overdrive*, December 15, 2016. https://www.overdriveonline.com/eyes-on-the-prize-the-market-for-pre-2000-trucks-in-light-of-eld-exemption/.

———. "E-Log Sticker Merits Easy Time at Weigh Station." *Overdrive*, February 12, 2017. https://www.overdriveonline.com/e-log-sticker-merits-easy-time-at-weigh-station/.

———. "One20 Launches the F-ELD." *Overdrive*, March 24, 2017. https://www.overdriveonline.com/one20-launches-the-f-eld/.

———. "Stoking the Log Fires: Hours Violations, Fleet Size and ELDs." *Overdrive*, July 6, 2017. https://www.overdriveonline.com/stoking-the-log-fires/.

———. "Finger on the ELD Mandate—Good Marketing or Well over the Line?" *Overdrive*, September 17, 2017. https://www.overdriveonline.com/finger-on-the-eld-mandate-good-marketing-or-well-over-the-line/.

———. "ELD Maker Launches Petition to Add 14-Hour Flexibility, Presents Data on Detention and Race-the-Clock Dynamics." *Overdrive*, November 19, 2017. https://www.overdriveonline.com/electronic-logging-devices/article/14893416/eld-maker-launches-petition-to-add-14-hour-flexibility-presents-data-on-detention-and-race-the-clock-dynamics.

———. "ELDs: What to Expect from Enforcement as the Mandate Descends." *Overdrive*, December 15, 2017. https://www.overdriveonline.com/elds-what-to-expect-from-enforcement-as-the-mandate-descends/.

———. "ELD Headaches: Dealing with Technical Glitches and Other Equipment Issues." *Overdrive*, March 27, 2018. https://www.overdriveonline.com/eld-headaches-dealing-with-technical-glitches-and-other-equipment-issues/.

———. "How to Blow the Whistle on Problem Shippers/Receivers When Delays Force Violations." *Overdrive*, March 29, 2018. https://www.overdriveonline.com/channel-19/article/14894114/how-to-blow-the-whistle-on-problem-shippersreceivers-when-delays-force-violations.

———. "So Much for the 'Hope for Leniency': One Owner-Operator's Hours Out-of-Service Experience in the ELD Era." *Overdrive*, May 13, 2018. https://www.overdriveonline.com/so-much-for-the-hope-for-leniency-one-owner-operators-hours-out-of-service-experience-in-the-eld-era/.

———. "Autonomous Revolution May Be Nigh, but It's Coming for Brokers Faster Than Operators." *Overdrive*, May 28, 2018. https://www.overdriveonline.com/autonomous-revolution-may-be-nigh-but-its-coming-for-brokers-faster-than-drivers/.

———. "ELDs Up the Ante on Parking." *Overdrive*, June 2, 2018. https://www.overdriveonline.com/electronic-logging-devices/article/14894459/how-parking-has-changed-in-light-of-the-eld-mandate.

———. "Guaranteed Pay Making Waves in Driver Compensation." *Overdrive*, July 2, 2018. https://www.overdriveonline.com/guaranteed-pay-making-waves-in-driver-compensation/.

———. "An ELD Positive, an Hours Negative, and More from Trucking's Present and Perhaps Stormy Future." *Overdrive*, July 29, 2018. https://www.overdriveonline.com/an-eld-positive-an-hours-negative-and-more-from-truckings-present-and-perhaps-stormy-future-in-expediting-panel/.

———. "FMCSA Director Discusses ELDs and Enforcement: Know What You're Dealing with at Roadside." *Overdrive*, August 24, 2018. https://www.overdriveonline.com/fmcsa-on-elds-and-enforcement-know-what-youre-dealing-with-at-roadside/.

———. "Fatigue's Fast Track—Dashcam Monitoring: Private Eyes, Watching You." *Overdrive*, December 18, 2018. https://www.overdriveonline.com/regulations/article/14893111/fatigues-fast-track-dashcam-monitoring-private-eyes-watching-you.

———. "Poll: Since the ELD Mandate, Have You Felt Pressure from Carriers/Customers to Violate Hours Regs More or Less Often?" *Overdrive*, March 26, 2019. https://www.overdriveonline .com/poll-since-the-eld-mandate-have-you-felt-pressure-from-carriers-customers-to-violate -hours-regs-more-or-less-often/.

———. "KeepTruckin to Leverage ELD Data for Freight Matching, Detention Pressure." *Over-drive*, April 17, 2019. https://www.overdriveonline.com/keeptruckin-to-leverage-eld-data -for-freight-matching-detention-pressure/.

———. "The Extent of Expanded Paid Parking Reservations after the ELD Mandate." *Over-drive*, July 22, 2021. https://www.overdriveonline.com/electronic-logging-devices/article /14894460/paid-parking-reservation-expansion-since-eld-mandate.

———. "The Not-So-Long Haul: The Operational Challenges to Owner-Ops after the ELD Mandate." *Overdrive*, July 22, 2021. https://www.overdriveonline.com/the-not-so-long-haul-the -operational-challenges-to-owner-ops-after-the-eld-mandate/.

DiMaggio, Paul J., and Walter W. Powell, "The Iron Cage Revisited: Institutional Isomorphism and Collective Rationality in Organizational Fields." *American Sociological Review* 48, no. 2 (1983): 147–60.

Dougherty, Conor. "Self-Driving Trucks May Be Closer Than They Appear." *New York Times*, November 13, 2017. https://www.nytimes.com/2017/11/13/business/self-driving-trucks .html.

Dunn, Naomi J., Jeffrey S. Hickman, Susan A. Soccolich, and Richard J. Hanowski. "Driver Detention Times in Commercial Motor Vehicle Operations." Federal Motor Carrier Safety Administration (report), December 2014. https://vtechworks.lib.vt.edu/bitstream/handle/10919 /55062/13-060-Detention-508C-Dec14.pdf.

Dwoskin, Elizabeth, and Mike Ramsey. "Car Makers Test Technology to Make You Pay Attention to the Road." *Wall Street Journal*, March 11, 2016. https://www.wsj.com/articles/car-tech-that -watches-how-you-drive-1457692201.

Edmondson, Catie. "'What Does a Trucker Look Like?' It's Changing, Amid a Big Shortage." *New York Times*, July 28, 2018. https://www.nytimes.com/2018/07/28/us/politics/trump -truck-driver-shortage.html.

"eFleetSuite—FMCSA 395 Subpart B Compliant Electronic Driver Logs Application." *Trimble IA Fleet Services*. https://store.isefleetservices.com/Articles.asp?ID=259.

Elish, Madeleine Clare. "Moral Crumple Zones: Cautionary Tales in Human-Robot Interaction." *Engaging Science, Technology, and Society* 5 (2019): 40–60.

"Emergency Declaration Information." Federal Motor Carrier Safety Administration, August 24, 2021. https://www.fmcsa.dot.gov/emergency-declarations#INFO.

Engel, Cynthia. "Competition Drives the Trucking Industry." *Monthly Labor Review* 121, no. 4 (1998): 34–41.

"Extended Detention Exception, The." *KeepTruckin*. https://go.keeptruckin.com/extended -detention-exception.

Ewick, Patricia, and Susan Silbey. "Conformity, Contestation, and Resistance: An Account of Legal Consciousness." *New England Law Review* 26: 731–49 (1991).

———. "Narrating Social Structure: Stories of Resistance to Legal Authority." *American Journal of Sociology* 108, no. 6 (2003): 1328–72.

Federal Motor Carrier Safety Administration. *Regulatory Impact Analysis of Electronic On-Board Recorders,* November 2006, FMCSA-2004-18940-0352.

———. *Proposed Rule: Electronic Logging Devices and Hours of Service Supporting Documents,* March 28, 2014, FMCSA-2010-0167-0485.

———. Hours of Service of Drivers: American Trucking Associations (ATA); Denial of Application for Exemption, 81 Federal Register 17240, March 28, 2016.

Federal Motor Carrier Safety Administration Office of Research and Analysis. "The Large Truck Crash Causation Study—Analysis Brief," July 2007. https://www.fmcsa.dot.gov/safety /research-and-analysis/large-truck-crash-causation-study-analysis-brief.

Ferreira, A., R. Cruz-Correia, L. Antunes, P. Farinha, E. Oliveira-Palhares, D.W. Chadwick, and A. Costa-Pereira. "How to Break Access Control in a Controlled Manner." In *Proceedings of the 19th IEEE International Symposium on Computer-Based Medical Systems*, 847–54. Los Alamitos, CA: IEEE, 2006.

Ferris, Robert. "There Is a Ton of People Who Still Don't Want to Ride in Self-Driving Cars, Says Survey." *CNBC*, August 24, 2017. https://www.cnbc.com/2017/08/24/consumers-still -anxious-about-autonomous-cars-says-gartner.html.

Fields, Mark. "Ford's Road to Full Autonomy." *LinkedIn*, April 16, 2016. https://www.linkedin .com/pulse/fords-road-full-autonomy-mark-fields/.

Fisher, Tyson. "What if the Transformers Had ELDs?" *Land Line Magazine* 43, no. 5 (July 2018): 52.

———. "Truck Fatalities Highest in 30 Years in First Full Year of ELD Mandate." *Land Line*, October 23, 2019. https://landline.media/truck-fatalities-highest-in-30-years-in-first-full-year -of-eld-mandate/.

"Five Trucking Companies Back EOBR Legislation." *Commercial Carrier Journal*, September 29, 2010. https://www.ccjdigital.com/business/article/14918249/5-trucking-companies-back -eobr-legislation.

Fonseca, Felicia, and Tom Krisher. "Self-Driving Vehicle Hits, Kills Walker in Arizona." *Columbus Dispatch*, March 20, 2018, A3–A4. http://www.pressreader.com/usa/the-columbus-dispatch /20180320/281513636687475.

Foucault, Michel. *Discipline & Punish: The Birth of the Prison*. New York: Vintage Books, 1977.

FOX59 Web Staff. "'Slow Roll' Rolls through Indy without Incident as Truckers Protest Regulations." *FOX59*, February 21, 2019. https://fox59.com/2019/02/21/police-warn-truckers-ahead -of-slow-roll-protest-around-i-465-thursday/.

Freedman, David H. "Self-Driving Trucks: 10 Breakthrough Technologies 2017." *MIT Technology Review*, February 22, 2017. Archived at https://web.archive.org/web/20200101172733 /https://www.technologyreview.com/s/603493/10-breakthrough-technologies-2017-self -driving-trucks/.

Freedman, Toby. "'Robot Man' Not Space Solution." *Miami News*, May 3, 1964, p. 10. Archived at https://perma.cc/JS9T-3GMK.

Frey, Carl Benedikt, and Michael A. Osborne. "The Future of Employment: How Susceptible Are Jobs to Computerisation?" *Technological Forecasting and Social Change* 114 (January 2017): 254–80.

Frydl, Kathleen. *The Drug Wars in America, 1940–1973*. Cambridge: Cambridge University Press, 2013.

Gabriel, Trip. "Alone on the Open Road: Truckers Feel Like 'Throwaway People.'" *New York Times*, May 22, 2017. https://www.nytimes.com/2017/05/22/us/trucking-jobs.html.

Gallagher, John. "What Lawmakers Intend to Do about Driver Detention." *Freight Waves*, June 4, 2020. https://www.freightwaves.com/news/lawmakers-prod-fmcsa-to-move-on-driver -detention.

Garba, Abubakar Bello, Jocelyn Armarego, David Murray, and William Kenworthy. "Review of the Information Security and Privacy Challenges in Bring Your Own Device (BYOD) Environments." *Journal of Information Privacy and Security* 11, no. 1 (2015): 38–54.

Garsten, Ed. "Sharp Growth in Autonomous Car Market Value Predicted but May Be Stalled by Rise in Consumer Fear." *Forbes*, August 13, 2018. https://www.forbes.com/sites/edgarsten /2018/08/13/sharp-growth-in-autonomous-car-market-value-predicted-but-may-be-stalled -by-rise-in-consumer-fear/#4e1144b4617c.

Gerzema, John, "Breaking down the Barriers of Self-Driving Cars." *Forbes*, November 16, 2017. https://www.forbes.com/sites/forbesagencycouncil/2017/11/16/breaking-down-the-barriers -of-self-driving-cars/?sh=6e7be4d22efl.

Gillespie, Tarleton. *Wired Shut: Copyright and the Shape of Digital Culture* Cambridge, MA: MIT Press, 2007.

Glaser, April. "Autonomous Tech Could Make Driving Semi-Trucks Even Less Fun." *Wired*, April 27, 2016. https://www.wired.com/2016/04/autonomous-tech-make-driving-semi -trucks-even-less-fun/.

Goffman, Erving. *The Presentation of Self in Everyday Life*. New York: Anchor Books, 1959.

———. "On Face-Work." In *Interaction Ritual: Essays on Face-to-Face Behavior*, 5–46. Garden City, NY: Anchor Books, 1967.

Goldenfein, Jake, Deirdre K. Mulligan, Helen Nissenbaum, and Wendy Ju. "Through the Handoff Lens: Competing Visions of Autonomous Futures." *Berkeley Technology Law Journal* 35, no. 3 (2020): 835–910.

Goode, Erica. "Nerve Damage to Brain Linked to Heavy Use of Ecstasy Drug." *New York Times*, October 30, 1998. https://www.nytimes.com/1998/10/30/us/nerve-damage-to-brain-linked -to-heavy-use-of-ecstasy-drug.html.

Goodman, J. David, and Matt Flegenheimer. "De Blasio's Vow to End Traffic Deaths Meets Reality of New York Streets." *New York Times*, February 14, 2014. https://www.nytimes.com/2014/02 /15/nyregion/vow-to-end-traffic-deaths-vs-reality-of-city-streets.html.

Gould, Jon B., and Scott Barclay. "Mind the Gap: The Place of Gap Studies in Sociolegal Scholarship." *Annual Review of Law and Social Science* 8 (2012): 323–35.

Gouldner, Alvin W. *Patterns of Industrial Bureaucracy*. New York: The Free Press, 1954.

"GPS Jamming: Out of Sight." *Economist*, July 27, 2013. https://www.economist.com/international /2013/07/27/out-of-sight.

Gray, Mary L., and Siddharth Suri. *Ghost Work: How to Stop Silicon Valley from Building a New Global Underclass*. Boston: Houghton Mifflin Harcourt, 2019.

Greenblatt, Alan. "The Racial History of the 'Grandfather Clause.'" *NPR*, October 22, 2013. https://www.npr.org/sections/codeswitch/2013/10/21/239081586/the-racial-history-of -the-grandfather-clause.

Greenlee, Eric T., Patricia R. DeLucia, and David C. Newton. "Driver Vigilance in Automated Vehicles: Hazard Detection Failures Are a Matter of Time." *Human Factors: The Journal of the Human Factors and Ergonomics Society* 60, no. 4 (2018): 465–76.

Gurley, Lauren Kaori. "Amazon Drivers Are Instructed to Drive Recklessly to Meet Delivery Quotas." *Vice*, May 6, 2021. https://www.vice.com/en/article/xgxx54/amazon-drivers-are -instructed-to-drive-recklessly-to-meet-delivery-quotas.

Hagemann, Ryan. "Senseless Government Rules Could Cripple the Robo-Car Revolution." *Wired*, May 1, 2017. https://www.wired.com/2017/05/senseless-government-rules-cripple-robo-car -revolution/.

Haggerty, Kevin D., and Richard V. Ericson. "The Surveillant Assemblage." *British Journal of Sociology* 51, no. 4 (2000): 605–22.

Halper, Evan. "The Driverless Revolution May Exact a Political Price" *Los Angeles Times*, November 21, 2017. https://www.latimes.com/politics/la-na-pol-self-driving-politics-20171121-story .html.

Hamilton, Shane. *Trucking Country: The Road to America's Wal-Mart Economy*. Princeton: Princeton University Press, 2008.

Hancock, P. A. "Some Pitfalls in the Promises of Automated and Autonomous Vehicles." *Ergonomics* 62, no. 4 (2019): 514–20.

Hansen, Daren. "Trucking Alliance Calls for ELD Use by Intrastate Drivers," J. J. Keller Encompass (blog), July 26, 2018. https://eld.kellerencompass.com/resource/blog/trucking-alliance-calls-for-eld-use-by-intra-state-drivers.

Hardy, Quentin. "How Urban Anonymity Disappears When All Data Is Tracked." *New York Times*, April 19, 2014. https://bits.blogs.nytimes.com/2014/04/19/how-urban-anonymity-disappears-when-all-data-is-tracked/.

Harris, Alexes. "Daunte Wright and the Grim Financial Incentive behind Traffic Stops." *Vox*, April 15, 2021. https://www.vox.com/first-person/22384104/daunte-wright-police-shooting-black-lives-matter-traffic-stops.

Hartzog, Woodrow, Gregory Conti, John Nelson, and Lisa A. Shay. "Insufficiently Automated Law Enforcement." *Michigan State Law Review* 2015, no. 5 (2015): 1763–96.

Heavy Duty Trucking Staff. "ZF Details Research on Passenger Monitoring System." *Truckinginfo*, October 16, 2018. https://www.truckinginfo.com/316828/zf-details-research-on-passenger-monitoring-system.

Heine, Max. "A Potential Role for Cameras and Wearables in Regulating Drowsy Driving." *Overdrive*, November 26, 2017. https://www.overdriveonline.com/fatigue-monitoring-eventual-replacement-for-hours-of-service/.

Hicks, Heather J. *The Culture of Soft Work: Labor, Gender, and Race in Postmodern American Narrative*. New York: Palgrave Macmillan, 2009.

Higgins, Tim. "Driverless Cars Still Handled by Humans-from Afar." *Wall Street Journal*, June 5, 2018. https://www.wsj.com/articles/who-does-a-driverless-car-call-when-it-needs-help-a-human-1528191000.

Hine, Christine. "Multi-Sited Ethnography as a Middle Range Methodology for Contemporary STS." *Science, Technology, & Human Values* 32, no. 6 (2007): 652–71.

Hirschman, Albert O. *Exit, Voice, and Loyalty: Responses to Decline in Firms, Organizations, and States*. Cambridge, MA: Harvard University Press, 1970.

Hull, Dana, Mark Bergen, and Polly Mosendz. "The Uber Crash Is the Nightmare the Driverless World Feared but Expected." *Bloomberg*, March 19, 2018. https://www.bloomberg.com/news/articles/2018-03-19/uber-crash-is-nightmare-the-driverless-world-feared-but-expected.

Humanyze. 2011. https://www.humanyze.com/.

Indiana State Police—Commercial Vehicle Enforcement Division. Facebook post, November 10, 2017. https://www.facebook.com/pg/ISPCVED/posts/.

International Transport Forum. *Managing the Transition to Driverless Road Freight Transport*. Paris: International Transport Forum, 2017. https://www.itf-oecd.org/sites/default/files/docs/managing-transition-driverless-road-freight-transport.pdf.

Jaillet, James. "The New Cost of e-Log Compliance and FMCSA's Denial of Small Business Exemption." *Overdrive*, December 13, 2015. https://www.overdriveonline.com/the-new-cost-of-e-log-compliance-and-fmcsas-denial-of-small-business-exemption/.

———. "Trucker Pay Has Plummeted in the Last 30 Years, Analyst Says." *Overdrive*, March 7, 2016. https://www.overdriveonline.com/trucker-pay-has-plummeted-in-the-last-30-years-analyst-stays/.

———. "Forecasters: Trucking's Autonomous Revolution Is Nigh, and It May Be Prickly." *Overdrive*, October 2, 2016. https://www.overdriveonline.com/forecasters-truckings-autonomous-revolution-is-nigh-and-it-may-be-prickly/.

———. "Study: Under ELDs, Crash Rates Flat but Unsafe Driving Violations Are Up." *Overdrive*, February 4, 2019. https://www.overdriveonline.com/electronic-logging-devices/article/14895792/study-under-elds-crash-rates-flat-but-unsafe-driving-violations-are-up.

———. "Protestors Reckon with Minimal 'Shutdown' and Protest Participation." *Overdrive*, April 20, 2019. https://www.overdriveonline.com/after-april-12-protests-fail-to-materialize-organizers-plan-smaller-relaunch-of-facebook-efforts/.

———. "Fatigue's Fast Track—Regulations: Are Fatigue-Monitoring Systems Destined to Upend Hours-Based Regulations?" *Overdrive*, September 23, 2019. https://www.overdriveonline.com/fatigues-fast-track-regulations-are-fatigue-monitoring-systems-destined-to-upend-hours-based-regulations/.

Jerolmack, Colin, and Shamus Khan. "Talk Is Cheap: Ethnography and the Attitudinal Fallacy." *Sociological Methods & Research* 43, no. 2 (2014): 178–209.

Jones, Jami. "Special Report: OOIDA Brief Details Driver Harassment by ATA Member Companies." *Land Line*, March 6, 2012. Archived at https://web.archive.org/web/20120307130122/http://www.landlinemag.com/Story.aspx?StoryID=23321.

Jonsson, Katrin, Jonny Holmström, and Kalle Lyytinen. "Turn to the Material: Remote Diagnostics Systems and New Forms of Boundary-Spanning." *Information and Organization* 19, no. 4 (2009): 233–52.

Juhlin, Oskar. "Traffic Behaviour as Social Interaction—Implications for the Design of Artificial Drivers." In *Proceedings of the 6th World Congress on Intelligent Transport Systems*, 8–12. Washington, DC: ITS America, 1999.

Kallinikos, Jannis. *The Consequences of Information: Institutional Implications of Technological Change*. Cheltenham, UK: Edward Elgar Publishing, 2007.

Kantor, Jodi, Karen Weise, and Grace Ashford. "The Amazon That Customers Don't See." *New York Times*, June 15, 2021. https://www.nytimes.com/interactive/2021/06/15/us/amazon-workers.html.

Kellogg, Katherine C., Melissa A. Valentine, and Angèle Christin. "Algorithms at Work: The New Contested Terrain of Control." *Academy of Management Annals* 14, no. 1 (2020): 366–410.

Kelly, Heather. "Amazon's Idea for Employee-Tracking Wearables Raises Concerns." *CNN Business*, February 2, 2018. https://money.cnn.com/2018/02/02/technology/amazon-employee-tracker/index.html.

Kerr, Ian. "Digital Locks and the Automation of Virtue." In *From "Radical Extremism" to "Balanced Copyright": Canadian Copyright and the Digital Agenda*, edited by Michael Geist. Toronto: Irwin Law, 2010.

Kerry, Cameron F., and Jack Karsten. "Gauging Investment in Self-Driving Cars." *Brookings*, October 16, 2017. https://www.brookings.edu/research/gauging-investment-in-self-driving-cars/.

Keynes, John Maynard. "Economic Possibilities for Our Grandchildren." In *Essays in Persuasion*, 358–73. London: Harcourt Brace, 1932.

Kilgour, Lauren. "Designed to Shame: Electronic Ankle Monitors and the Politics of Carceral Technology." PhD diss., Cornell University, 2021.

"Knights on the Highway." Promotional film produced by the Jam Handy Organization, 1938. Archived at https://archive.org/details/Knights01938.

Körber, Moritz, Andrea Cingel, Markus Zimmermann, and Klaus Bengler. "Vigilance Decrement and Passive Fatigue Caused by Monotony in Automated Driving." *Procedia Manufacturing* 3 (2015): 2403–9.

Korosec, Kirsten. "This Driverless Truck Startup Is Putting Human Drivers to Work." *Fortune*, February 28, 2017. https://fortune.com/2017/02/28/starsky-self-driving-truck-startup/.

———. "Uber's Self-Driving Trucks Division Is Dead, Long Live Uber Self-Driving Cars." *TechCrunch*, July 30, 2018. https://techcrunch.com/2018/07/30/ubers-self-driving-trucks-division-is-dead-long-live-uber-self-driving-cars/.

Krisher, Tom. "The Toughest Challenge for Self-Driving Cars? Human Drivers." *AP News*, May 11, 2017. https://apnews.com/65e1c800ff0c4532abbdabc134e5f9b8.

Lee, Timothy B. "Legislation Would Mandate Driver-Monitoring Tech in Every Car." *Ars Technica*, April 27, 2021. https://arstechnica.com/tech-policy/2021/04/legislation-would-mandate-driver-monitoring-tech-in-every-car/.

Levin, Alan, and Ryan Beene. "Tesla Model X in California Crash Sped Up Prior to Impact." *Bloomberg*, June 7, 2018. https://www.bloomberg.com/news/articles/2018-06-07/tesla-model-x-in-california-crash-sped-up-seconds-before-impact.

Levy, Karen, and Solon Barocas. "Refractive Surveillance: Monitoring Customers to Manage Workers." *International Journal of Communication* 12 (2018): 1166–88.

Levy, Karen, and Michael Franklin. "Driving Regulation: Using Topic Models to Examine Political Contention in the United States Trucking Industry." *Social Science Computer Review* 32, no. 2 (2014) :182–94.

Lin, Tzuoo-Ding, Paul P. Jovanis, and Chun-Zin Yang. "Modeling the Safety of Truck Driver Service Hours Using Time-Dependent Logistic Regression." *Transportation Research Record* iss. 1407 (1993): 1–10.

Lipton, Eric. "How $225,000 Can Help Secure a Pollution Loophole at Trump's E.P.A." *New York Times*, February 15, 2018. https://www.nytimes.com/2018/02/15/us/politics/epa-pollution-loophole-glider-trucks.html.

———. "'Super Polluting' Trucks Receive Loophole on Pruitt's Last Day." *New York Times*, July 6, 2018, https://www.nytimes.com/2018/07/06/us/glider-trucks-loophole-pruitt.html.

Long, Mindy. "Truck Stops Step Up to Combat Parking Crisis." *Fleet Owner*, December 31, 2019. https://www.fleetowner.com/safety/article/21119313/truck-stops-step-up-to-combat-parking-crisis.

Mackworth, N. H. "The Breakdown of Vigilance during Prolonged Visual Search." *Quarterly Journal of Experimental Psychology* 1, no. 1 (1948): 6–21.

Madrigal, Alexis C. "Could Self-Driving Trucks Be Good for Truckers?" *Atlantic*, February 1, 2018. https://www.theatlantic.com/technology/archive/2018/02/uber-says-its-self-driving-trucks-will-be-good-for-truckers/551879/.

Mann, Steve, Jason Nolan, and Barry Wellman. "Sousveillance: Inventing and Using Wearable Computing Devices for Data Collection in Surveillance Environments." *Surveillance and Society* 1, no. 3 (2003): 331–55.

Marchant, Gary E. "The Pacing Problem." In *The Growing Gap between Emerging Technologies and Legal-Ethical Oversight*, edited by Gary E. Marchant, Braden R. Allenby, and Joseph R. Herkert, 199–205. New York: Springer, 2011.

Marcus, George E. "Ethnography in/of the World System: The Emergence of Multi-Sited Ethnography." *Annual Review of Anthropology* 24 (1995): 95–117.

Markoff, John. "Google's Next Phase in Driverless Cars: No Steering Wheel or Brake Pedals." *New York Times*, May 27, 2014. https://www.nytimes.com/2014/05/28/technology/googles-next-phase-in-driverless-cars-no-brakes-or-steering-wheel.html.

Martin, Aaron K., Rosamunde E. Van Brakel, and Daniel J. Bernhard. "Understanding Resistance to Digital Surveillance: Towards a Multi-Disciplinary, Multi-Actor Framework." *Surveillance and Society* 6, no. 3 (2009): 213–32.

Marx, Gary T. "A Tack in the Shoe: Neutralizing and Resisting the New Surveillance." *Journal of Social Issues* 59, no. 2 (2003): 369–90.

Mateescu, Alexandra, and Madeleine Clare Elish. *AI in Context: The Labor of Integrating New Technologies*. Data & Society Research Institute (report), 2019. https://datasociety.net/wp-content/uploads/2019/01/DataandSociety_AIinContext.pdf.

Matwyshyn, Andrea M. "The Internet of Bodies." *William & Mary Law Review* 61, no. 1 (2019): 77–167.

McFarland, Matt. "A Self-Driving Truck Just Hauled 51,744 Cans of Budweiser on a Colorado Highway." *CNN Business*, October 25, 2016. https://money.cnn.com/2016/10/25/technology /otto-budweiser-self-driving-truck/index.html.

McKinsey Center for Future Mobility. "The Future of Mobility is at Our Doorstep." McKinsey Compendium 2019/2020. https://www.mckinsey.com/~/media/McKinsey/Industries /Automotive%20and%20Assembly/Our%20Insights/The%20future%20of%20mobility%20 is%20at%20our%20doorstep/The-future-of-mobility-is-at-our-doorstep.ashx.

Mello, John E., and C. Shane Hunt. "Developing a Theoretical Framework for Research into Driver Control Practices in the Trucking Industry." *Transportation Journal* 48, no. 4 (2009): 20–39.

Menegus, Bryan. "Amazon Imagines Future Employee Surveillance with Patent Application for AR Goggles." *Gizmodo*, August 2, 2018. https://gizmodo.com/amazon-imagines-future -employee-surveillance-with-paten-1828051062.

Merton, Robert K. Foreword to *The Technological Society*, by Jacques Ellul. New York: Vintage, 1964.

———. "Three Fragments from a Sociologist's Notebooks: Establishing the Phenomenon, Specified Ignorance, and Strategic Research Materials." *Annual Review of Sociology* 13, no. 1 (1987): 1–29.

Metz, Cade, and Kate Conger. "Uber, After Years of Trying, Is Handing Off Its Self-Driving Car Project." *New York Times*, December 7, 2020. https://www.nytimes.com/2020/12/07 /technology/uber-self-driving-car-project.html.

Meyer, John W., and Brian Rowan. "Institutionalized Organizations: Formal Structure as Myth and Ceremony." *American Journal of Sociology* 83, no. 2 (1977): 340–63.

Miller, Daniel. *Car Cultures*. New York: Routledge, 2001.

Miller, Ross. "AP's 'Robot Journalists' Are Writing Their Own Stories Now." *Verge*, January 29, 2015. https://www.theverge.com/2015/1/29/7939067/ap-journalism-automation-robots -financial-reporting.

Molotch, Harvey. *Against Security: How We Go Wrong at Airports, Subways, and Other Sites of Ambiguous Danger*. Princeton: Princeton University Press, 2014.

Monahan, Torin. "Counter-Surveillance as Political Intervention?" *Social Semiotics* 16, no. 4 (2006): 515–34.

———. "The Right to Hide? Anti-Surveillance Camouflage and The Aestheticization of Resistance." *Communication and Critical/Cultural Studies* 12, no. 2 (2015): 159–78.

Monahan, Torin, and Jill A. Fisher. "Benefits of 'Observer Effects': Lessons from the Field." *Qualitative Research* 10, no. 3 (2010): 357–76.

Moradi, Pegah, and Karen Levy. "The Future of Work in the Age of AI: Displacement or Risk-Shifting?" In *The Oxford Handbook of Ethics of AI*, edited by Markus D. Dubber, Frank Pasquale, and Sunit Das, 271–88. Oxford: Oxford University Press, 2020.

Morris, Michael W., Kwok Leung, Daniel Ames, and Brian Lickel. "Views from Inside and Outside: Integrating Emic and Etic Insights about Culture and Justice Judgment." *Academy of Management Review* 24, no. 4 (1999): 781–96.

Morse, Jack. "Microsoft Waters Down 'Productivity Score' Surveillance Tool After Backlash." *Mashable*, December 1, 2020. https://mashable.com/article/microsoft-365-productivity -score-workplace-surveillance-backlash.

Morse, Janice M., Michael Barrett, Maria Mayan, Karin Olson, and Jude Spiers. "Verification Strategies for Establishing Reliability and Validity in Qualitative Research." *International Journal of Qualitative Methods* 1, no. 2 (2002): 13–22.

Mulligan, Christina M. "Perfect Enforcement of Law: When to Limit and When to Use Technology." *Richmond Journal of Law and Technology* 14, no. 4 (2008): 1–49.

Murphy, Brett. "Asleep at the Wheel." *USA Today*, December 28, 2017. https://www.usatoday.com/pages/interactives/news/rigged-asleep-at-the-wheel/.

Nader, Laura. "Up the Anthropologist: Perspectives Gained from Studying Up." In *Reinventing Anthropology*, edited by Dell Hymes, 284–311. New York: Pantheon Books, 1972.

Napier, Brian. "Working to Rule—A Breach of the Contract of Employment?" *Industrial Law Journal* 1, no. 3 (1972): 125–34.

Narayanan, Arvind. "What Happened to the Crypto Dream? Part 2," *IEEE Security & Privacy* 11, no. 3 (2013): 0068–0071.

National Academies of Science, Engineering, and Medicine. *Commercial Motor Vehicle Driver Fatigue, Long-Term Health, and Highway Safety: Research Needs*. Washington, DC: The National Academies Press, 2016. https://www.nap.edu/catalog/21921/commercial-motor-vehicle-driver-fatigue-long-term-health-and-highway-safety.

National Highway Traffic Safety Administration. "Consumer Advisory: NHTSA Deems 'Autopilot Buddy' Product Unsafe," June 19, 2018. https://www.nhtsa.gov/press-releases/consumer-advisory-nhtsa-deems-autopilot-buddy-product-unsafe.

———. "Traffic Safety Facts: 2019 Data," May 2021. https://crashstats.nhtsa.dot.gov/Api/Public/ViewPublication/813110.

National Transportation Safety Board. "Accident Report: Collision Between a Car Operating with Automated Vehicle Control Systems and a Tractor-Semitrailer Truck Near Williston, Florida." NTSB/HAR-17/02 PB2017–102600, May 7, 2016. https://www.ntsb.gov/investigations/AccidentReports/Reports/HAR1702.pdf.

Naughton, Keith. "Most Americans Wary of Self-Driving Tech, Don't Want Robo Cars." *Bloomberg*, September 6, 2017. https://www.bloomberg.com/news/articles/2017-09-06/most-americans-wary-of-self-driving-tech-don-t-want-robo-cars.

Nerad, Jack. "Trucking Industry Seeks to Avoid 'Robot Apocalypse.'" *Trucks.com*, May 9, 2018. https://www.trucks.com/2018/05/09/trucking-industry-robot-apocalypse/.

Newell, Bryce. *Police Visibility: Privacy, Surveillance, and the False Promise of Body-Worn Cameras*. Oakland, CA: University of California Press, 2021.

Nosowitz, Dan. "The Long White Line: The Mental and Physical Effects of Long-Haul Trucking." *Pacific Standard*, June 14, 2017. https://psmag.com/social-justice/the-long-white-line-the-mental-and-physical-effects-of-long-haul-trucking.

Nyholm, Sven, and Jilles Smids. "Automated Cars Meet Human Drivers: Responsible Human-Robot Coordination and the Ethics of Mixed Traffic." *Ethics and Information Technology* 22, no. 4 (2018): 335–44.

"Occupational Employment and Wages, May 2020: 53–3032 Heavy and Tractor-Trailer Truck Drivers." United States Bureau of Labor Statistics. Archived at https://web.archive.org/web/20210713164113/https://www.bls.gov/oes/current/oes533032.htm.

O'Kane, Sean. "Tesla Starts Using In-Car Camera for Autopilot Driver Monitoring." *Verge*, May 27, 2021. https://www.theverge.com/2021/5/27/22457430/tesla-in-car-camera-driver-monitoring-system.

Omnitracs. *Omnitracs ELD Driver Retention Model*. https://www.omnitracs.com/sites/default/files/files/2018-10/ELD_Driver_Retention_Model_08_16_Brochure_web.pdf.

Onuoha, Mimi. "When Proof Is Not Enough." *FiveThirtyEight*, July 1, 2020. https://fivethirtyeight.com/features/when-proof-is-not-enough/.

Orlikowski, Wanda J. "The Duality of Technology: Rethinking the Concept of Technology in Organizations." *Organization Science* 3, no. 3 (1992): 398–427.

Ouellet, Lawrence J. *Pedal to the Metal: The Work Lives of Truckers.* Philadelphia: Temple University Press, 1994.

Overdrive Staff. "PrePass Expands Bypass Capabilities, Introduces ELD." *Overdrive,* October 23, 2017. https://www.overdriveonline.com/prepass-expands-bypass-capabilities-introduces-eld/.

———. "Daimler Trucks Building Facility Dedicated to Autonomous Research, Development." *Overdrive,* June 13, 2018. https://www.overdriveonline.com/business/article/14894536/daimer-trucks-to-open-new-research-development-center.

———. "Hours of Service Violation Rate Cut in Half under ELD Mandate, FMCSA Says." *Overdrive,* June 25, 2018. https://www.overdriveonline.com/electronic-logging-devices/article/14894600/hours-of-service-violation-rate-cut-in-half-under-eld-mandate-fmcsa-says.

———. "What ELD Exodus? Major Broker's Anecdotal Take on Capacity Constraints under Mandate." *Overdrive,* August 8, 2018. https://www.overdriveonline.com/what-eld-exodus-major-brokers-anecdotal-take-on-capacity-constraints-under-mandate/.

———. "Data Sharing Practices among Leading Providers of ELDs to Owner-Operators." *Overdrive,* June 17, 2019. https://www.overdriveonline.com/electronic-logging-devices/article/14896412/data-sharing-practices-among-leading-providers-of-elds-to-owner-operators.

———. "ELD Data Handling: 'Privacy Is Paramount,' but Practices Vary." *Overdrive,* June 17, 2019. https://www.overdriveonline.com/electronic-logging-devices/article/14896447/eld-data-handling-privacy-is-paramount-but-practices-vary.

———. "More Hopes and Hazards: ELD Data Aggregation for Advocacy, Predictive Maintenance, Other Uses." *Overdrive,* February 20, 2020. https://www.overdriveonline.com/electronic-logging-devices/article/14896514/more-hopes-and-hazards-eld-data-aggregation-for-advocacy-predictive-maintenance-other-uses.

———. "A Gold Rush for ELD Data." *Overdrive,* July 21, 2021. https://www.overdriveonline.com/electronic-logging-devices/article/14896446/a-gold-rush-for-eld-data.

Owen, Diane S. *Deregulation in the Trucking Industry.* Washington, DC: Federal Trade Commission Bureau of Economics, 1988.

Perlman, David, Dan Bogard, Alex Epstein, Antonio Santalucia, and Anita Kim. "Review of the Federal Motor Carrier Safety Regulations for Automated Commercial Vehicles: Preliminary Assessment of Interpretation and Enforcement Challenges, Questions, and Gaps." Summary report prepared for Intelligent Transportation Systems Joint Program Office and Federal Motor Carrier Safety Administration, Report no. FMCSA-RRT-17–013, March 2018. https://rosap.ntl.bts.gov/view/dot/35426.

Pethokoukis, James. "What the Story of ATMs and Bank Tellers Reveals About the 'Rise of the Robots' and Jobs." American Enterprise Institute, June 6, 2016. https://www.aei.org/economics/what-atms-bank-tellers-rise-robots-and-jobs/.

Pilon, Mary. "Surviving the Long Haul." *New Republic,* July 12, 2016. https://newrepublic.com/article/135000/surviving-long-haul.

Pinch, Trevor J., and Wiebe E. Bijker. "The Social Construction of Facts and Artefacts: Or How the Sociology of Science and the Sociology of Technology Might Benefit Each Other." *Social Studies of Science* 14, no. 3 (1984): 399–441.

Polanyi, Michael. *The Tacit Dimension.* Chicago: University of Chicago Press, 1966.

"Police Investigate Truck Driver's Facebook Brag about 20 Hours of Driving." *CDL Life,* November 18, 2015. https://cdllife.com/2015/police-investigate-truck-drivers-facebook-brag-about-20-hours-of-driving/.

Rai, Supriya. "Tech Mahindra's Moodometer Gauges Employees' Mood to Uplift Work Environment." *Dataquest,* August 10, 2018. https://www.dqindia.com/tech-mahindras-moodometer-gauges-employees-mood-uplift-work-environment/.

"Real-time Driver Fatigue Monitoring System: *Guardian*." Seeing Machines. https://www
.seeingmachines.com/guardian/guardian/.

Regalado, Antonio. "When Machines Do Your Job." *MIT Technology Review*, July 11, 2012. https://
www.technologyreview.com/s/428429/when-machines-do-your-job/.

Reich, Robert B. *Saving Capitalism: For the Many, Not the Few*. New York: Alfred A. Knopf, 2016.

Rich, Michael L. "Should We Make Crime Impossible?" *Harvard Journal of Law and Public Policy*
36, no. 2 (2013): 795–848.

Roca, Cristina, and Dieter Holger. "Drawn by the Salary, Women Flock to Trucking." *Wall Street
Journal*, October 14, 2019. https://www.wsj.com/articles/drawn-by-the-salary-women-flock
-to-trucking-11571045406.

Rolland, Knut H., and Eric Monteiro. "Balancing the Local and the Global in Infrastructural
Information Systems." *Information Society* 18, no. 2 (2002): 87–100.

Rosenblat, Alex, and Luke Stark. "Algorithmic Labor and Information Asymmetries: A Case Study
of Uber's Drivers." *International Journal of Communication* 10 (2016): 3758–84.

Rubin, Ashley T. "The Consequences of Prisoners' Micro-Resistance." *Law & Social Inquiry* 42,
no. 1 (2017): 138–62.

Sagberg, Fridulv, Selpi, Giulio Francesco Bianchi Piccinini, and Johan Engström. "A Review of
Research on Driving Styles and Road Safety." *Human Factors: The Journal of the Human Factors
and Ergonomics Society* 57, no. 7 (2015): 1248–75.

Said, Carolyn. "Robot Cars May Kill Jobs, but Will They Create Them Too?" *San Francisco Chronicle*, December 8, 2017. https://www.sfchronicle.com/news/article/Robot-cars-may-kill-jobs
-but-will-they-create-12410820.php.

Saxby, Dyani J., Gerald Matthews, Joel S. Warm, Edward M. Hitchcock, and Catherine Neu-
bauer. "Active and Passive Fatigue in Simulated Driving: Discriminating Styles of Workload
Regulation and Their Safety Impacts." *Journal of Experimental Psychology: Applied* 19, no. 4
(2013): 287–300.

Schremmer, Mark. "New Gadgets Move Us Closer to a World of 'RoboTruckers.'" *Land Line*, January 11, 2017. https://landline.media/new-gadgets-move-us-closer-to-a-world-of-robotruckers/.

———. "Be a Minute Late, and There Will Be Hell to Pay." *Land Line*, May 9, 2018. https://landline
.media/be-a-minute-late-and-there-will-be-hell-to-pay/.

———. "ELD Mandate Hasn't Reduced Crashes, Study Reveals." *Land Line*, February 5, 2019.
https://landline.media/eld-mandate-hasnt-reduced-crashes-study-reveals/.

———. "Large Fleets Reduce Driver Count; Turnover Rate Hits 96%." *Land Line*, December 23,
2019. https://landline.media/large-fleets-reduce-driver-count-turnover-rate-hits-96/.

———. "It's Time to Treat the Root Cause of Driver Retention Problem." *Land Line*, November 24,
2021. https://landline.media/its-time-to-treat-the-root-cause-of-driver-retention-problem/.

———. "OOIDA on Improving Retention: 'Put It on the Paycheck.'" *Land Line*, December 17,
2021. https://landline.media/ooida-on-improving-retention-put-it-on-the-paycheck/.

Scism, Leslie. "America's Truckers Embrace Big Brother after Costing Insurers Millions." *Wall
Street Journal*, June 4, 2017. https://www.wsj.com/articles/americas-truckers-embrace-big
-brother-after-costing-insurers-millions-1496577601.

Scott, Alex, Andrew Balthrop, and Jason W. Miller. "Unintended Responses to IT-Enabled Monitoring: The Case of the Electronic Logging Device Mandate." *Journal of Operations Management* 67, no. 2 (2021): 152–81.

Scott, James C. *Weapons of the Weak: Everyday Forms of Peasant Resistance*. New Haven: Yale
University Press, 1987.

———. *Seeing Like a State: How Certain Schemes to Improve the Human Condition Have Failed*.
New Haven: Yale University Press, 1998.

Seale, Clive. "Quality in Qualitative Research." *Qualitative Inquiry* 5, no. 4 (1999): 465–78.

Seltz-Axmacher, Stefan. "The End of Starsky Robotics." Medium post, March 19, 2020. https://medium.com/starsky-robotics-blog/the-end-of-starsky-robotics-acb8a6a8a5f5.

Sewell, Graham. "The Discipline of Teams: The Control of Team-Based Industrial Work through Electronic and Peer Surveillance." *Administrative Science Quarterly* 43, no. 2 (1998): 397–428.

Shattell, Mona, Yorghos Apostolopoulos, Sevil Sönmez, and Mary Griffin. "Occupational Stressors and the Mental Health of Truckers." *Issues in Mental Health Nursing* 31, no. 9 (2010): 561–68.

Sheridan, Thomas B. "Big Brother as Driver: New Demands and Problems for the Man at the Wheel." *Human Factors: The Journal of the Human Factors and Ergonomics Society* 12, no. 1 (1970): 95–101.

Short, Jeffrey. "White Paper: Analysis of Truck Driver Age Demographics Across Two Decades." American Transportation Research Institute (report), December 2014. https://truckingresearch.org/wp-content/uploads/2014/12/Analysis-of-Truck-Driver-Age-Demographics-FINAL-12-2014.pdf.

Short, Jeffrey, and Dan Murray. "Identifying Autonomous Vehicle Technology Impacts on the Trucking Industry." American Transportation Research Institute (report), November 2016. https://truckingresearch.org/wp-content/uploads/2016/11/ATRI-Autonomous-Vehicle-Impacts-11-2016.pdf.

Sieber, W. Karl, Cynthia F. Robinson, Jan Birdsey, Guang X. Chen, Edward M. Hitchcock, Jennifer E. Lincoln, Akinori Nakata, and Marie H. Sweeney. "Obesity and Other Risk Factors: The National Survey of U.S. Long-Haul Truck Driver Health and Injury." *American Journal of Industrial Medicine* 57, no. 6 (2014): 615–26.

Smith, Alan, and Forrest Lucas. "A New Electronic Logging Rule Could Drive Independent Truckers off the Road." *Hill*, October 17, 2017. https://thehill.com/opinion/energy-environment/355563-a-new-electronic-logging-rule-could-drive-independent-truckers-off.

Smith, Bryant Walker. "A Legal Perspective on Three Misconceptions in Vehicle Automation." In *Road Vehicle Automation*, edited by Gereon Meyer and Sven Beiker, 85–91. New York: Springer, 2014.

Smith, Sean, and Patrick Harris. "Truck Driver Job-Related Injuries in Overdrive." United States Department of Labor Blog, August 17, 2017. Archived at https://web.archive.org/web/20170427031745/https:/blog.dol.gov/2016/08/17/truck-driver-job-related-injuries-in-overdrive.

Speltz, Erin, and Dan Murray. "Driver Detention Impacts on Safety and Productivity." American Transportation Research Institute (report), September 2019. https://truckingresearch.org/wp-content/uploads/2019/09/ATRI-Detention-Impacts-09-2019.pdf.

Stanley, Karlyn D., Michelle Grisé, and James M. Anderson. "Autonomous Vehicles and the Future of Auto Insurance." RAND (report), 2020. https://www.rand.org/pubs/research_reports/RRA878-1.html.

Strand, Ginger. *Killer on the Road: Violence and the American Interstate*. Austin, TX: University of Texas Press, 2014.

Stratford, Dale, Tedd V. Ellerbrock, J. Keith Akins, and Heather L. Hall. "Highway Cowboys, Old Hands, and Christian Truckers: Risk Behavior for Human Immunodeficiency Virus Infection among Long-Haul Truckers in Florida." *Social Science & Medicine* 50, no. 5 (2000): 737–49.

Stromberg, Joseph. "This Is the First Licensed Self-Driving Truck. There Will Be Many More." *Vox*, May 6, 2015. https://www.vox.com/2014/6/3/5775482/why-trucks-will-drive-themselves-before-cars-do.

Stumpf, Rob. "Tesla Admits Current 'Full Self-Driving Beta' Will Always Be a Level 2 System: Emails." *Drive*, March 8, 2021. https://www.thedrive.com/tech/39647/tesla-admits-current-full-self-driving-beta-will-always-be-a-level-2-system-emails.

Sullivan, Gail. "Trucker in Tracy Morgan Crash Hadn't Slept for More Than 24 Hours." *Washington Post*, June 10, 2014. https://www.washingtonpost.com/news/morning-mix/wp/2014/06/10/trucker-in-tracy-morgan-crash-hadnt-slept-for-more-than-24-hours/.

"Survey Finds Larger Truck Fleets Ready for ELD Mandate; Small Firms, Not So Much." *DC Velocity*, September 14, 2016. https://www.dcvelocity.com/articles/28213-survey-finds-larger-truck-fleets-ready-for-eld-mandate-small-firms-not-so-much.

Tanner, David. "'Why Aren't You Rolling?'" *Land Line Magazine*, December 2013: 32–33.

Taub, Eric A. "Sleepy Behind the Wheel? Some Cars Can Tell." *New York Times*, March 16, 2017. https://www.nytimes.com/2017/03/16/automobiles/wheels/drowsy-driving-technology.html.

"Teamsters Union Perspective on Automation Featured in *Rolling Stone*." International Brotherhood of Teamsters, January 2, 2018. https://teamster.org/2018/01/teamsters-union-perspective-automation-featured-rolling-stone/.

Templeton, Brad. "Starsky Robotics Shuts Down and Worries Everybody Else Will Also Fail in Robotic Trucks." *Forbes*, April 2, 2020. https://www.forbes.com/sites/bradtempleton/2020/04/02/starsky-robotics-shuts-down-and-worries-everybody-else-will-also-fail-in-robotic-trucks/.

Tennant, Chris, and Jack Stilgoe. "The Attachments of 'Autonomous' Vehicles." *Social Studies of Science* 51, no. 6 (2021): 846–70, p. 849.

"Things Your Drivers Don't Know About Roadside Inspections." *Oil Prophets: A Quarterly Publication by the Petroleum & Convenience Marketers of Alabama* (Summer 2018): 18–19. https://emflipbooks.com/flipbooks/PCMA/OilProhets_Magazine/Summer18/book/18/.

Thompson, Iain. "Feds Arrest Rogue Trucker after GPS Jamming Borks New Jersey Airport Test." *Register*, August 12, 2013. https://www.theregister.com/2013/08/12/feds_arrest_rogue_trucker_after_gps_jamming_disrupts_newark_airport/.

"Trucking Company Tells Tired Truck Drivers: Don't Stop When You're Tired." *Trucking Watchdog*, July 6, 2016. https://www.truckingwatchdog.com/2016/07/06/trucking-company-tells-drivers-dont-stop-when-youre-tired/.

Uber Advanced Technologies Group. "The Future of Trucking." Medium post, November 14, 2017. Archived at https://web.archive.org/web/20180210230753/https://medium.com/@UberATG/the-future-of-trucking-292202cb42a2.

———. "The Future of Trucking: Mixed Fleets, Transfer Hubs, and More Opportunity for Truck Drivers." Medium post, February 1, 2018. Archived at https://web.archive.org/web/20180201192005/https://medium.com/@UberATG/the-future-of-trucking-b3d2ea0d2db9.

United States Department of Transportation. "U.S. Department of Transportation Declares Illinois Long-Haul Truck Driver to Be an Imminent Hazard to Public Safety," February 12, 2014. https://www.transportation.gov/briefing-room/us-department-transportation-declares-illinois-long-haul-truck-driver-be-imminent-0.

———. "Analysis of Driver Critical Reason and Years of Driving Experience in Large Truck Crashes," January 2017. https://rosap.ntl.bts.gov/view/dot/31693.

United States Department of Transportation Office of Inspector General. "Enhanced FAA Oversight Could Reduce Hazards Associated with Increased Use of Flight Deck Automation." Report No. AV-2016–013, January 7, 2016. https://www.oig.dot.gov/sites/default/files/FAA%20Flight%20Decek%20Automation_Final%20Report%5E1-7-16.pdf.

———. "Estimates Show Commercial Driver Detention Increases Crash Risks and Costs, but Current Data Limit Further Analysis." Report No. ST2018019, January 31, 2018. https://www.oig.dot.gov/sites/default/files/FMCSA%20Driver%20Detention%20Final%20Report.pdf.

Vanderwerp, Dave. "It's Not Just Tesla: All Other Driver-Assist Systems Work without Drivers, Too." *Car and Driver*, August 11, 2021. https://www.caranddriver.com/news/a37260363/driver-assist-systems-tested/.

Vaughan, Diane. *Controlling Unlawful Organizational Behavior: Social Structure and Corporate Misconduct*. Chicago: University of Chicago Press, 1985.

Viscelli, Steve. *The Big Rig: Trucking and the Decline of the American Dream*. Oakland, CA: University of California Press, 2016.

———. "Truck Stop: How One of America's Steadiest Jobs Turned into One of Its Most Grueling." *Atlantic*, May 10, 2016. https://www.theatlantic.com/business/archive/2016/05/truck-stop/481926/.

———. "Driverless? Autonomous Trucks and the Future of the American Trucker." Report of the Center for Labor Research and Education, University of California, Berkeley, and Working Partnerships USA, September 2018, 16–17. http://driverlessreport.org/files/driverless.pdf.

———. "A Pragmatic View of How ELDs Are Changing Trucking." *Fleet Owner*, December 26, 2018. https://www.fleetowner.com/technology/telematics/article/21703331/a-pragmatic-view-of-how-elds-are-changing-trucking.

Vitak, Jessica, and Michael Zimmer. "Workers' Attitudes toward Increased Surveillance during and after the Covid-19 Pandemic." *Items* (blog), Social Science Research Council, September 16, 2021. https://items.ssrc.org/covid-19-and-the-social-sciences/covid-19-fieldnotes/workers-attitudes-toward-increased-surveillance-during-and-after-the-covid-19-pandemic/.

Vogelpohl, Tobias, Matthias Kühn, Thomas Hummel, and Mark Vollrath. "Asleep at the Automated Wheel—Sleepiness and Fatigue during Highly Automated Driving." *Accident Analysis & Prevention* 126 (May 2019): 70–84.

"Volvo Cars CEO Urges Governments and Car Industry to Share Safety-Related Traffic Data." Volvo Cars Global Media Newsroom, April 3, 2017. https://www.media.volvocars.com/global/en-gb/media/pressreleases/207164/volvo-cars-ceo-urges-governments-and-car-industry-to-share-safety-related-traffic-data.

Watson, Rip. "Driver Demographics Begin to Shift." *Transport Topics*, December 11, 2014. https://www.ttnews.com/articles/driver-demographics-begin-shift.

Wells, Helen, and David Wills. "Individualism and Identity Resistance to Speed Cameras in the UK." *Surveillance and Society* 6, no. 3 (2009): 259–74.

Wilhelm, Alex. "Bring on the Blogging Robots." *TechCrunch*, July 1, 2014. https://techcrunch.com/2014/07/01/bring-on-the-blogging-robots/.

Willcocks, Leslie. "Robo-Apocalypse Cancelled? Reframing the Automation and Future of Work Debate." *Journal of Information Technology* 35, no. 4 (2020): 286–302.

Willis, Paul. *Learning to Labor: How Working Class Kids Get Working Class Jobs*. New York: Columbia University Press, 1981.

Wolski, Chris. "Telematics Gamification Emphasizes Fun over 'Big Brother.'" *Automotive Fleet*, October 21, 2015. https://www.automotive-fleet.com/156349/telematics-gamification-emphasizes-fun-over-big-brother.

Wong, Joon Ian. "A Fleet of Trucks Just Drove Themselves across Europe." *Quartz*, April 6, 2016. https://qz.com/656104/a-fleet-of-trucks-just-drove-themselves-across-europe/.

Yankelevich, Aleksandr, R. V. Rikard, Travis Kadylak, Michael J. Hall, Elizabeth A. Mack, John P. Verboncoeur, and Sheila R. Cotten. "Preparing the Workforce for Automated Vehicles." Report submitted to The American Center for Mobility, July 30, 2018. https://comartsci.msu.edu/sites/default/files/documents/MSU-TTI-Preparing-Workforce-for-AVs-and-Truck-Platooning-Reports%20.pdf.

Zittrain, Jonathan L. *The Future of the Internet and How to Stop It*. New Haven: Yale University Press, 2008.

Zuboff, Shoshana. *In the Age of the Smart Machine: The Future of Work and Power*. New York: Basic Books, 1988.

INDEX

Page numbers in italic type indicate illustrations

Printed and bound by CPI Group (UK) Ltd, Croydon, CR0 4YY

25/03/2025

14647346-0001